HOLT SCIENCE & TECHNOLOGY

Introduction to Matter

HOLT, RINEHART AND WINSTON

A Harcourt Classroom Education Company

Austin · New York · Orlando · Atlanta · San Francisco · Boston · Dallas · Toronto · London

Acknowledgments

Chapter Writers

Christie Borgford, Ph.D.
Professor of Chemistry
University of Alabama
Birmingham, Alabama

Andrew Champagne
Former Physics Teacher
Ashland High School
Ashland, Massachusetts

Mapi Cuevas, Ph.D.
Professor of Chemistry
Santa Fe Community College
Gainesville, Florida

Leila Dumas
Former Physics Teacher
LBJ Science Academy
Austin, Texas

William G. Lamb, Ph.D.
Science Teacher and Dept. Chair
Oregon Episcopal School
Portland, Oregon

Sally Ann Vonderbrink, Ph.D.
Chemistry Teacher
St. Xavier High School
Cincinnati, Ohio

Lab Writers

Phillip G. Bunce
Former Physics Teacher
Bowie High School
Austin, Texas

Kenneth E. Creese
Science Teacher
White Mountain Junior High School
Rock Springs, Wyoming

William G. Lamb, Ph.D.
Science Teacher and Dept. Chair
Oregon Episcopal School
Portland, Oregon

Alyson Mike
Science Teacher
East Valley Middle School
East Helena, Montana

Joseph W. Price
Science Teacher and Dept. Chair
H. M. Browne Junior High School
Washington, D.C.

Denice Lee Sandefur
Science Teacher and Dept. Chair
Nucla High School
Nucla, Colorado

John Spadafino
Mathematics and Physics Teacher
Hackensack High School
Hackensack, New Jersey

Walter Woolbaugh
Science Teacher
Manhattan Junior High School
Manhattan, Montana

Academic Reviewers

Paul R. Berman, Ph.D.
Professor of Physics
University of Michigan
Ann Arbor, Michigan

Russell M. Brengelman, Ph.D.
Professor of Physics
Morehead State University
Morehead, Kentucky

John A. Brockhaus, Ph.D.
Director, Mapping, Charting and Geodesy Program
Department of Geography and Environmental Engineering
United States Military Academy
West Point, New York

Walter Bron, Ph.D.
Professor of Physics
University of California
Irvine, California

Andrew J. Davis, Ph.D.
Manager, ACE Science Center
Department of Physics
California Institute of Technology
Pasadena, California

Peter E. Demmin, Ed.D.
Former Science Teacher and Department Chair
Amherst Central High School
Amherst, New York

Roger Falcone, Ph.D.
Professor of Physics and Department Chair
University of California
Berkeley, California

Cassandra A. Fraser, Ph.D.
Assistant Professor of Chemistry
University of Virginia
Charlottesville, Virginia

L. John Gagliardi, Ph.D.
Associate Professor of Physics and Department Chair
Rutgers University
Camden, New Jersey

Gabriele F. Giuliani, Ph.D.
Professor of Physics
Purdue University
West Lafayette, Indiana

Roy W. Hann, Jr., Ph.D.
Professor of Civil Engineering
Texas A&M University
College Station, Texas

John L. Hubisz, Ph.D.
Professor of Physics
North Carolina State University
Raleigh, North Carolina

Samuel P. Kounaves, Ph.D.
Professor of Chemistry
Tufts University
Medford, Massachusetts

Karol Lang, Ph.D.
Associate Professor of Physics
The University of Texas
Austin, Texas

Gloria Langer, Ph.D.
Professor of Physics
University of Colorado
Boulder, Colorado

Phillip LaRoe
Professor
Helena College of Technology
Helena, Montana

Joseph A. McClure, Ph.D.
Associate Professor of Physics
Georgetown University
Washington, D.C.

LaMoine L. Motz, Ph.D.
Coordinator of Science Education
Department of Learning Services
Oakland County Schools
Waterford, Michigan

R. Thomas Myers, Ph.D.
Professor of Chemistry, Emeritus
Kent State University
Kent, Ohio

Hillary Clement Olson, Ph.D.
Research Associate
Institute for Geophysics
The University of Texas
Austin, Texas

David P. Richardson, Ph.D.
Professor of Chemistry
Thompson Chemical Laboratory
Williams College
Williamstown, Massachusetts

John Rigden, Ph.D.
Director of Special Projects
American Institute of Physics
Colchester, Vermont

Acknowledgments (cont.)

Peter Sheridan, Ph.D.
Professor of Chemistry
Colgate University
Hamilton, New York

Vederaman Sriraman,
Ph.D.
Associate Professor of Technology
Southwest Texas State University
San Marcos, Texas

Jack B. Swift, Ph.D.
Professor of Physics
The University of Texas
Austin, Texas

Atiq Syed, Ph.D.
Master Instructor of Mathematics and Science
Texas State Technical College
Harlingen, Texas

Leonard Taylor, Ph.D.
Professor Emeritus
Department of Electrical Engineering
University of Maryland
College Park, Maryland

Virginia L. Trimble, Ph.D.
Professor of Physics and Astronomy
University of California
Irvine, California

Martin VanDyke, Ph.D.
Professor of Chemistry, Emeritus
Front Range Community College
Westminster, Colorado

Gabriela Waschewsky,
Ph.D.
Science and Math Teacher
Emery High School
Emeryville, California

Safety Reviewer

Jack A. Gerlovich, Ph.D.
Associate Professor
School of Education
Drake University
Des Moines, Iowa

Teacher Reviewers

Barry L. Bishop
Science Teacher and Dept. Chair
San Rafael Junior High School
Ferron, Utah

Paul Boyle
Science Teacher
Perry Heights Middle School
Evansville, Indiana

Kenneth Creese
Science Teacher
White Mountain Junior High School
Rock Springs, Wyoming

Vicky Farland
Science Teacher and Dept. Chair
Centennial Middle School
Yuma, Arizona

Rebecca Ferguson
Science Teacher
North Ridge Middle School
North Richland Hills, Texas

Laura Fleet
Science Teacher
Alice B. Landrum Middle School
Ponte Vedra Beach, Florida

Jennifer Ford
Science Teacher and Dept. Chair
North Ridge Middle School
North Richland Hills, Texas

Susan Gorman
Science Teacher
North Ridge Middle School
North Richland Hills, Texas

C. John Graves
Science Teacher
Monforton Middle School
Bozeman, Montana

Dennis Hanson
Science Teacher and Dept. Chair
Big Bear Middle School
Big Bear Lake, California

David A. Harris
Science Teacher and Dept. Chair
The Thacher School
Ojai, California

Norman E. Holcomb
Science Teacher
Marion Local Schools
Maria Stein, Ohio

Kenneth J. Horn
Science Teacher and Dept. Chair
Fallston Middle School
Fallston, Maryland

Tracy Jahn
Science Teacher
Berkshire Junior-Senior High School
Canaan, New York

Kerry A. Johnson
Science Teacher
Isbell Middle School
Santa Paula, California

Drew E. Kirian
Science Teacher
Solon Middle School
Solon, Ohio

Harriet Knops
Science Teacher and Dept. Chair
Rolling Hills Middle School
El Dorado, California

Scott Mandel, Ph.D.
Director and Educational Consultant
Teachers Helping Teachers
Los Angeles, California

Thomas Manerchia
Former Science Teacher
Archmere Academy
Claymont, Delaware

Edith McAlanis
Science Teacher and Dept. Chair
Socorro Middle School
El Paso, Texas

Kevin McCurdy, Ph.D.
Science Teacher
Elmwood Junior High School
Rogers, Arkansas

Alyson Mike
Science Teacher
East Valley Middle School
East Helena, Montana

Donna Norwood
Science Teacher and Dept. Chair
Monroe Middle School
Charlotte, North Carolina

Joseph W. Price
Science Teacher and Dept. Chair
H. M. Browne Junior High School
Washington, D.C.

Terry J. Rakes
Science Teacher
Elmwood Junior High School
Rogers, Arkansas

Beth Richards
Science Teacher
North Middle School
Crystal Lake, Illinois

Elizabeth J. Rustad
Science Teacher
Crane Middle School
Yuma, Arizona

Rodney A. Sandefur
Science Teacher
Naturita Middle School
Naturita, Colorado

Helen Schiller
Science Teacher
Northwood Middle School
Taylors, South Carolina

Bert J. Sherwood
Science Teacher
Socorro Middle School
El Paso, Texas

Patricia McFarlane Soto
Science Teacher and Dept. Chair
G. W. Carver Middle School
Miami, Florida

David M. Sparks
Science Teacher
Redwater Junior High School
Redwater, Texas

Larry Tackett
Science Teacher and Dept. Chair
Andrew Jackson Middle School
Cross Lanes, West Virginia

Elsie N. Waynes
Science Teacher and Dept. Chair
R. H. Terrell Junior High School
Washington, D.C.

Sharon L. Woolf
Science Teacher
Langston Hughes Middle School
Reston, Virginia

Alexis S. Wright
Middle School Science Coordinator
Rye Country Day School
Rye, New York

Lee Yassinski
Science Teacher
Sun Valley Middle School
Sun Valley, California

John Zambo
Science Teacher
Elizabeth Ustach Middle School
Modesto, California

K Introduction to Matter

Skills Development

Process Skills

QuickLabs

Chapter Labs

Skills Development *(continued)*

Research and Critical Thinking Skills

Apply

Feature Articles

Connections

Mathematics

To the Student

This book was created to make your science experience interesting, exciting, and fun!

Go for It!

Science is a process of discovery, a trek into the unknown. The skills you develop using *Holt Science & Technology*—such as observing, experimenting, and explaining observations and ideas— are the skills you will need for the future. There is a universe of exploration and discovery awaiting those who accept the challenges of science.

Science & Technology

You see the interaction between science and technology every day. Science makes technology possible. On the other hand, some of the products of technology, such as computers, are used to make further scientific discoveries. In fact, much of the scientific work that is done today has become so technically complicated and expensive that no one person can do it entirely alone. But make no mistake, the creative ideas for even the most highly technical and expensive scientific work still come from individuals.

Activities and Labs

The activities and labs in this book will allow you to make some basic but important scientific discoveries on your own. You can even do some exploring on your own at home! Here's your chance to use your imagination and curiosity as you investigate your world.

Keep a ScienceLog

In this book, you will be asked to keep a type of journal called a ScienceLog to record your thoughts, observations, experiments, and conclusions. As you develop your ScienceLog, you will see your own ideas taking shape over time. You'll have a written record of how your ideas have changed as you learn about and explore interesting topics in science.

Know "What You'll Do"

The "What You'll Do" list at the beginning of each section is your built-in guide to what you need to learn in each chapter. When you can answer the questions in the Section Review and Chapter Review, you know you are ready for a test.

Check Out the Internet

You will see this logo throughout the book. You'll be using *sci*LINKS as your gateway to the Internet. Once you log on to *sci*LINKS using your computer's Internet link, type in the *sci*LINKS address. When asked for the keyword code, type in the keyword for that topic. A wealth of resources is now at your disposal to help you learn more about that topic.

In addition to *sci*LINKS you can log on to some other great resources to go with your text. The addresses shown below will take you to the home page of each site.

This textbook contains the following on-line resources to help you make the most of your science experience.

go.hrw.com	**SCiLINKS** NSTA	**Smithsonian Institution®** Internet Connections	**CNNfyi.com**
Visit **go.hrw.com** for extra help and study aids matched to your textbook. Just type in the keyword HST HOME.	Visit **www.scilinks.org** to find resources specific to topics in your textbook. Keywords appear throughout your book to take you further.	Visit **www.si.edu/hrw** for specifically chosen on-line materials from one of our nation's premier science museums.	Visit **www.cnnfyi.com** for late-breaking news and current events stories selected just for you.

The Properties of Matter

Pre-Reading Questions

1. What is matter?
2. What is the difference between a physical property and a chemical property?
3. What is the difference between a physical change and a chemical change?

NICE ICE

You've seen water in many forms: steam rising from a kettle, dew collecting on grass, and tiny crystals of frost forming on the windows in winter. But no matter what its form, water is still water. In this chapter, you'll learn more about the many different properties of matter, such as water. You'll also learn about changes in matter that take place all around you.

SACK SECRETS

In this activity, you will test your skills in determining the identity of an object based on its properties.

Procedure

1. You and two or three of your classmates will receive a **sealed paper sack** containing a **mystery object.** Do not open the sack!

2. For 5 minutes, make as many observations as you can about the object. You may touch, smell, or listen to the object through the sack; shake the sack; and so on. Be sure to record your observations.

Analysis

3. At the end of 5 minutes, discuss your findings with your partners.

4. In your ScienceLog, list the object's properties. Make a conclusion about the object's identity.

5. Share your observations, your list of properties, and your conclusion with the class. Now you are ready to open the sack.

6. Did you properly identify the object? If so, how? If not, why not? Write your answers in your ScienceLog. Share them with the class.

Terms to Learn

matter
volume
meniscus
mass
gravity
weight
newton
inertia

What You'll Do

- Name the two properties of all matter.
- Describe how volume and mass are measured.
- Compare mass and weight.
- Explain the relationship between mass and inertia.

Space Case

1. Crumple a **piece of paper,** and fit it tightly in the bottom of a **cup** so that it won't fall out.

2. Turn the cup upside down. Lower the cup straight down into a **large beaker or bucket** half-filled with **water** until the cup is all the way underwater.

3. Lift the cup straight out of the water. Turn the cup upright and observe the paper. Record your observations in your ScienceLog.

4. Now punch a small hole in the bottom of the cup with the point of a **pencil.** Repeat steps 2 and 3.

5. How do these results show that air has volume? Record your explanation in your ScienceLog.

What Is Matter?

Here's a strange question: What do you have in common with a toaster?

Give up? Okay, here's another question: What do you have in common with a steaming bowl of soup or a bright neon sign?

You are probably thinking these are trick questions. After all, it is hard to imagine that a human—you—has anything in common with a kitchen appliance, some hot soup, or a glowing neon sign.

Everything Is Made of Matter

From a scientific point of view you have at least one characteristic in common with these things. You, the toaster, the bowl, the soup, the steam, the glass tubing, and the glowing gas are all made of matter. But what is matter exactly? If so many different kinds of things are made of matter, you might expect the definition of the word *matter* to be complicated. But it is really quite simple. **Matter** is anything that has volume and mass.

Matter Has Volume

All matter takes up space. The amount of space taken up, or occupied, by an object is known as the object's **volume.** The sun, shown in **Figure 1,** has volume because it takes up space at the center of our solar system. Your fingernails, the Statue of Liberty, the continent of Africa, and a cloud all have volume. And because these things have volume, they cannot share the same space at the same time. Even the tiniest speck of dust takes up space, and there's no way another speck of dust can fit into that space without somehow bumping the first speck out of the way. Try the QuickLab on this page to see for yourself that matter takes up space.

Figure 1 *The volume of the sun is about 1,000,000 (1 million) times larger than the volume of the Earth.*

Liquid Volume   Lake Erie, the smallest of the Great Lakes, has a volume of approximately 483,000,000,000,000 (483 trillion) liters of water. Can you imagine that much liquid? Well, think of a 2 liter bottle of soda. The water in Lake Erie could fill more than 241 trillion of those bottles. That's a lot of water! On a smaller scale, a can of soda has a volume of only 355 milliliters, which is approximately one-third of a liter. You can read the volume printed on the soda can. Or you can check the volume by pouring the soda into a large measuring cup from your kitchen, as shown in **Figure 2.**

Figure 2   *If the measurement is accurate, the volume measured should be the same as the volume printed on the can.*

Measuring the Volume of Liquids   In your science class, you'll probably use a graduated cylinder to measure the volume of liquids. Keep in mind that the surface of a liquid in a graduated cylinder is not flat. The curve that you see at the liquid's surface has a special name—the **meniscus** (muh NIS kuhs). When you measure the volume of a liquid, you must look at the bottom of the meniscus, as shown in **Figure 3.** (A liquid in any container, including a measuring cup or a large beaker, has a meniscus. The meniscus is just too flat to see in a wider container.)

Liters (L) and milliliters (mL) are the units used most often to express the volume of liquids. The volume of any amount of liquid, from one raindrop to a can of soda to an entire ocean, can be expressed in these units.

Figure 3   *To measure volume correctly, read the scale at the lowest part of the meniscus (as indicated) at eye level.*

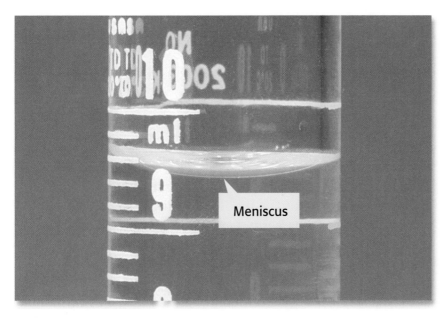

Meniscus

BRAIN FOOD

The volume of a typical raindrop is approximately 0.09 mL, which means that it would take almost 4,000 raindrops to fill a soda can.

LabBook

How would you measure the volume of this strangely shaped object? To find out, turn to page 134 in the LabBook.

Solid Volume The volume of any solid object is expressed in cubic units. *Cubic* means "having three dimensions." One cubic unit, a cubic meter, is shown in **Figure 4**. In science, cubic meters (m³) and cubic centimeters (cm³) are the units most often used to express the volume of solid items. The *3* in these unit abbreviations shows that three quantities were multiplied to get the final result. For a rectangular object, these three quantities are length, width, and height. Try this for yourself in the MathBreak at left.

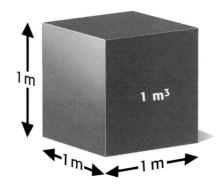

Figure 4 *A cubic meter has a height of 1 m, a length of 1 m, and a width of 1 m, so its volume is 1 m × 1 m × 1 m = 1 m³.*

Comparing Solid and Liquid Volumes Suppose you want to determine whether the volume of an ice cube is equal to the volume of water that is left when the ice cube melts. Because 1 mL is equal to 1 cm³, you can express the volume of the water in cubic centimeters and compare it with the volume of the ice cube. The volume of any liquid can be expressed in cubic units in this way. (However, in SI, volumes of solids are never expressed in liters or milliliters.)

Measuring the Volume of Gases How do you measure the volume of a gas? You can't hold a ruler up to a gas, and you can't pour a gas into a graduated cylinder. So it's impossible, right? Wrong! A gas expands to fill its container, so if you know the volume of the container the gas is in, then you know the volume of the gas.

Matter Has Mass

Another characteristic of all matter is mass. **Mass** is the amount of matter that something is made of. For example, the Earth is made of a very large amount of matter and therefore has a large mass. A peanut is made of a much smaller amount of matter and thus has a smaller mass. Remember, even something as small as a speck of dust is made of matter and therefore has mass.

Is a Puppy Like a Bowling Ball? An object's mass can be changed only by changing the amount of matter in the object. Consider the bowling ball shown in **Figure 5**. Its mass is constant because the amount of matter in the bowling ball never changes (unless you use a sledgehammer to remove a chunk of it!). Now consider the puppy. Does its mass remain constant? No, because the puppy is growing. If you measured the puppy's mass next year or even next week, you'd find that it had increased. That's because more matter—more puppy—would be present.

Figure 5 *The mass of the bowling ball does not change. The mass of the puppy increases as more matter is added—that is, as the puppy grows.*

The Difference Between Mass and Weight

Weight is different from mass. To understand this difference, you must first understand gravity. **Gravity** is a force of attraction between objects that is due to their masses. This attraction causes objects to exert a pull on other objects. Because all matter has mass, all matter experiences gravity. The amount of attraction between objects depends on two things—the masses of the objects and the distance between them, as shown in **Figure 6.**

Figure 6   How Mass and Distance Affect Gravity Between Objects

a Gravitational force (represented by the width of the arrows) is large between objects with large masses that are close together.

b Gravitational force is smaller between objects with smaller masses that are close together than between objects with large masses that are close together (as shown in **a**).

c An increase in distance reduces gravitational force between two objects. Therefore, gravitational force between objects with large masses (such as those in **a**) is less if they are far apart.

The Properties of Matter   **7**

May the Force Be with You Gravitational force is experienced by all objects in the universe all the time. But the ordinary objects you see every day have masses so small (relative to, say, planets) that their attraction toward each other is hard to detect. Therefore, the gravitational force experienced by objects with small masses is very slight. However, the Earth's mass is so large that the gravitational force between objects, such as our atmosphere or the space shuttle, and the Earth is great. Gravitational force is what keeps you and everything else on Earth from floating into space.

So What About Weight? A measure of the gravitational force exerted on an object is called **weight.** Consider the brick in **Figure 7.** The brick has mass. The Earth also has mass. Therefore, the brick and the Earth are attracted to each other. A force is exerted on the brick because of its attraction to the Earth. The weight of the brick is a measure of this gravitational force.

Now look at the sponge in Figure 7. The sponge is the same size as the brick, but its mass is much less. Therefore, the sponge's attraction toward the Earth is not as great, and the gravitational force on the sponge is not as great. Thus, the *weight* of the sponge is less than the *weight* of the brick.

At a Distance The attraction between objects decreases as the distance between them increases. As a result, the gravitational force exerted on objects also decreases as the distance increases. For this reason, a brick floating in space would weigh less than it does resting on Earth's surface. However, the brick's mass would stay the same.

Figure 7 *This brick and sponge may be the same size, but their masses, and therefore their weights, are quite different.*

Massive Confusion Back on Earth, the gravitational force exerted on an object is about the same everywhere, so an object's weight is also about the same everywhere. Because mass and weight remain constant everywhere on Earth, the terms *mass* and *weight* are often used as though they mean the same thing. But using the terms interchangeably can lead to confusion. So remember, weight depends on mass, but weight is not the same thing as mass.

Measuring Mass and Weight

The SI unit of mass is the kilogram (kg), but mass is often expressed in grams (g) and milligrams (mg) as well. These units can be used to express the mass of any object, from a single cell in your body to the entire solar system. Weight is a measure of gravitational force and must be expressed in units of force. The SI unit of force is the **newton (N).** So weight is expressed in newtons.

A newton is approximately equal to the weight of a 100 g mass on Earth. So if you know the mass of an object, you can calculate its weight on Earth. Conversely, if you know the weight of an object on Earth, you can determine its mass. **Figure 8** summarizes the differences between mass and weight.

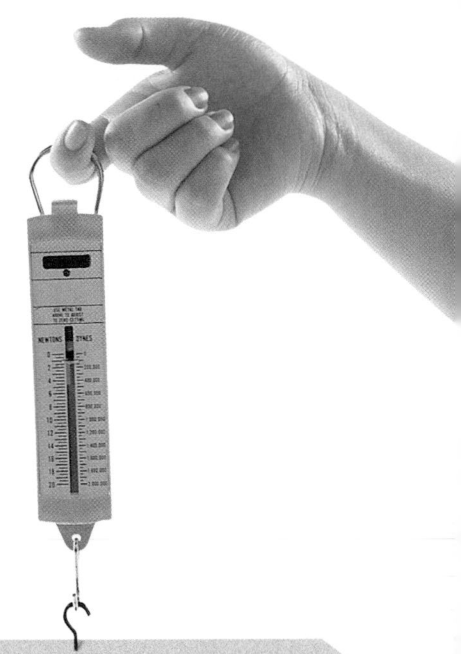

Figure 8   Differences Between Mass and Weight

Mass is . . .

■ a measure of the amount of matter in an object.

■ always constant for an object no matter where the object is in the universe.

■ measured with a balance (shown below).

■ expressed in kilograms (kg), grams (g), and milligrams (mg).

Weight is . . .

■ a measure of the gravitational force on an object.

■ varied depending on where the object is in relation to the Earth (or any other large body in the universe).

■ measured with a spring scale (shown above).

■ expressed in newtons (N).

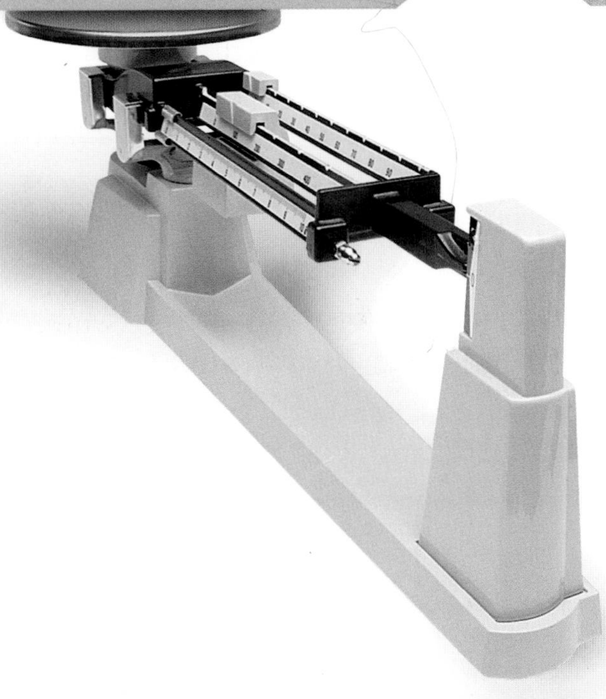

✓ Self-Check

If all of your school books combined have a mass of 3 kg, what is their total weight in newtons? Remember that 1 kg = 1,000 g. *(See page 168 to check your answer.)*

Mass, Weight, and Bathroom Scales

Ordinary bathroom scales are spring scales. Many scales available today show a reading in both pounds (a common, though not SI, unit of weight) and kilograms. How does such a reading contribute to the confusion between mass and weight?

Mass Is a Measure of Inertia

Imagine trying to kick a soccer ball that has the mass of a bowling ball. It would be painful! The reason has to do with inertia (in UHR shuh). **Inertia** is the tendency of all objects to resist any change in motion. Because of inertia, an object at rest will remain at rest until something causes it to move. Likewise, a moving object continues to move at the same speed and in the same direction unless something acts on it to change its speed or direction.

Mass is a measure of inertia because an object with a large mass is harder to start in motion and harder to stop than an object with a smaller mass. This is because the object with the large mass has greater inertia. For example, imagine that you are going to push a grocery cart that has only one potato in it. No problem, right? But suppose the grocery cart is filled with potatoes, as in **Figure 9.** Now the total mass—and the inertia—of the cart full of potatoes is much greater. It will be harder to get the cart moving and harder to stop it once it is moving.

Figure 9 *Why is a cartload of potatoes harder to get moving than a single potato? Because of inertia, that's why!*

📑 internet**connect**

SC*i*LINKS.
NSTA

TOPIC: What Is Matter?
GO TO: www.scilinks.org
*sci*LINKS NUMBER: HSTP030

SECTION REVIEW

1. What are the two properties of all matter?

2. How is volume measured? How is mass measured?

3. **Analyzing Relationships** Do objects with large masses always have large weights? Explain your reasoning.

Terms to Learn

physical property   physical change
density   chemical change
chemical property

What You'll Do

- ◆ Give examples of matter's different properties.
- ◆ Describe how density is used to identify different substances.
- ◆ Compare physical and chemical properties.
- ◆ Explain what happens to matter during physical and chemical changes.

Describing Matter

Have you ever heard of the game called "20 Questions"? In this game, your goal is to determine the identity of an object that another person is thinking of by asking questions about the object. The other person can respond with only a "yes" or a "no." If you can identify the object after asking 20 or fewer questions, you win! If you still can't figure out the object's identity after asking 20 questions, you may not be asking the right kinds of questions.

What kinds of questions should you ask? You might find it helpful to ask questions about the properties of the object. Knowing the properties of an object can help you determine the object's identity, as shown below.

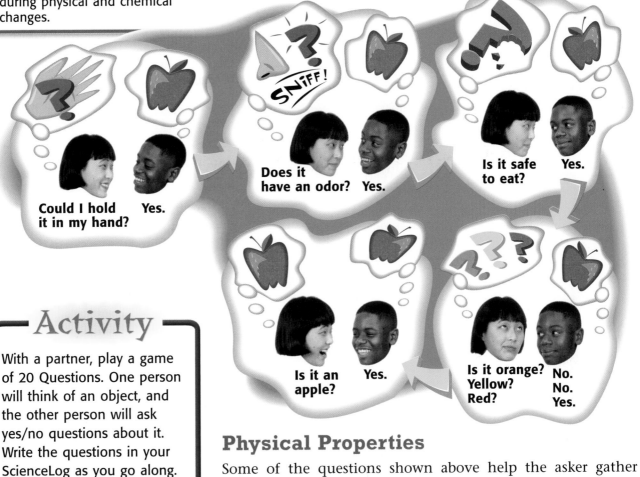

Activity

With a partner, play a game of 20 Questions. One person will think of an object, and the other person will ask yes/no questions about it. Write the questions in your ScienceLog as you go along. Put a check mark next to the questions asked about physical properties. When the object is identified or when the 20 questions are up, switch roles. Good luck!

Physical Properties

Some of the questions shown above help the asker gather information about *color* (Is it orange?), *odor* (Does it have an odor?), and *mass* and *volume* (Could I hold it in my hand?). Each of these properties is a physical property of matter. A **physical property** of matter can be observed or measured without changing the identity of the matter. For example, you don't have to change what the apple is made of to see that it is red or to hold it in your hand.

The Properties of Matter   **11**

Physical Properties Identify Matter You rely on physical properties all the time. For example, physical properties help you determine whether your socks are clean (odor), whether you can fit all your books into your backpack (volume), or whether your shirt matches your pants (color). The table below lists some more physical properties that are useful in describing or identifying matter.

More Physical Properties		
Physical property	**Definition**	**Example**
Thermal conductivity	the ability to transfer thermal energy from one area to another	Plastic foam is a poor conductor, so hot chocolate in a plastic-foam cup will not burn your hand.
State	the physical form in which a substance exists, such as a solid, liquid, or gas	Ice is water in its solid state.
Malleability (MAL ee uh BIL uh tee)	the ability to be pounded into thin sheets	Aluminum can be rolled or pounded into sheets to make foil.
Ductility (duhk TIL uh tee)	the ability to be drawn or pulled into a wire	Copper is often used to make wiring.
Solubility (SAHL yoo BIL uh tee)	the ability to dissolve in another substance	Sugar dissolves in water.
Density	mass per unit volume	Lead is used to make sinkers for fishing line because lead is more dense than water.

Spotlight on Density Density is a very helpful property when you need to distinguish different substances. Look at the definition of density in the table above—mass per unit volume. If you think back to what you learned in Section 1, you can define density in other terms: **density** is the amount of matter in a given space, or volume, as shown in **Figure 10.**

Figure 10 *A golf ball is more dense than a table-tennis ball because the golf ball contains more matter in a similar volume.*

To find an object's density (D), first measure its mass (m) and volume (V). Then use the following equation:

$$D = \frac{m}{V}$$

Units for density are expressed using a mass unit divided by a volume unit, such as g/cm³, g/mL, kg/m³, and kg/L.

Using Density to Identify Substances Density is a useful property for identifying substances for two reasons. First, the density of a particular substance is always the same at a given pressure and temperature. For example, the helium in a huge airship has a density of 0.0001663 g/cm³ at 20°C and normal atmospheric pressure. You can calculate the density of any other sample of helium at that same temperature and pressure—even the helium in a small balloon—and you will get 0.0001663 g/cm³. Second, the density of one substance is usually different from that of another substance. Check out the table below to see how density varies among substances.

Densities of Common Substances*			
Substance	Density (g/cm³)	Substance	Density (g/cm³)
Helium (gas)	0.0001663	Copper (solid)	8.96
Oxygen (gas)	0.001331	Silver (solid)	10.50
Water (liquid)	1.00	Lead (solid)	11.35
Iron pyrite (solid)	5.02	Mercury (liquid)	13.55
Zinc (solid)	7.13	Gold (solid)	19.32

at 20°C and normal atmospheric pressure

Mass = 96.6 g
Volume = 5.0 cm³

Figure 11 Is this gold or fool's gold?

The nugget in **Figure 11** looks like gold. But is it? It might be fool's gold instead. Fool's gold is another name for iron pyrite (PIE RIET), a mineral that looks like gold. How can you tell what the nugget really is? You could compare the density of the nugget with the known densities for gold and iron pyrite at the same temperature and pressure. By comparing densities, you'd know whether you've struck gold or you've been fooled!

Figure 12 *The yellow liquid is the least dense, and the green liquid is the densest.*

Liquid Layers What do you think causes the liquid in **Figure 12** to look the way it does? Is it magic? Is it trick photography? No, it's differences in density! There are actually four different liquids in the jar. Each liquid has a different density. Because of these differences in density, the liquids do not mix together but instead separate into layers, with the densest layer on the bottom and the least dense layer on top. The order in which the layers separate helps you determine how the densities of the liquids compare with one another.

The Density Challenge Imagine that you could put a lid on the jar in the picture and shake up the liquids. Would the different liquids mix together so that the four colors would blend into one interesting color? Maybe for a minute or two. But if the liquids are not soluble in one another, they would start to separate, and eventually you'd end up with the same four layers.

The same thing happens when you mix oil and vinegar to make salad dressing. When the layers separate, the oil is on top. But what do you think would happen if you added more oil? What if you added so much oil that there was several times as much oil as there was vinegar? Surely the oil would get so heavy that it would sink below the vinegar, right? Wrong! No matter how much oil you have, it will always be less dense than the vinegar, so it will always rise to the top. The same is true of the four liquids shown in Figure 12. Even if you add more yellow liquid than all of the other liquids combined, all of the yellow liquid will rise to the top. That's because density does not depend on how much of a substance you have.

Density and Grease Separators

The grease separator shown here is a kitchen device that cooks use to collect the best meat juices for making gravies. Based on what you know about density, describe how a grease separator works. Be sure to explain why the spout is at the bottom.

SECTION REVIEW

1. List three physical properties of water.

2. Why does a golf ball feel heavier than a table-tennis ball?

3. Describe how you can determine the relative densities of liquids.

4. **Applying Concepts** How could you determine that a coin is not pure silver?

Chemical Properties

Physical properties are not the only properties that describe matter. **Chemical properties** describe a substance based on its ability to change into a new substance with different properties. For example, a piece of wood can be burned to create new substances (ash and smoke) with properties different from the original piece of wood. Wood has the chemical property of *flammability*—the ability to burn. A substance that does not burn, such as gold, has the chemical property of nonflammability. Other common chemical properties include reactivity with oxygen, reactivity with acid, and reactivity with water. (The word *reactivity* just means that when two substances get together, something can happen.)

Observing Chemical Properties Chemical properties can be observed with your senses. However, chemical properties aren't as easy to observe as physical properties. For example, you can observe the flammability of wood only while the wood is burning. Likewise, you can observe the nonflammability of gold only when you try to burn it and it won't burn. But a substance always has its chemical properties. A piece of wood is flammable even when it's not burning.

Some Chemical Properties of Car Maintenance Look at the old car shown in **Figure 13**. Its owner calls it Rust Bucket. Why has this car rusted so badly while some other cars the same age remain in great shape? Knowing about chemical properties can help answer this question.

Most car bodies are made from steel, which is mostly iron. Iron has many desirable physical properties, including strength, malleability, and a high melting point. Iron also has many desirable chemical properties, including nonreactivity with oil and gasoline. All in all, steel is a good material to use for car bodies. It's not perfect, however, as you can probably tell from the car shown here.

Paint doesn't react with oxygen, so it provides a barrier between oxygen and the iron in the steel.

This hole started as a small chip in the paint. The chip exposed the iron in the car's body to oxygen. The iron rusted and eventually crumbled away.

Figure 13 Rust Bucket
One unfavorable chemical property of iron is its reactivity with oxygen. When iron is exposed to oxygen, it rusts.

This bumper is rust free because it is coated with a barrier of chromium, which is nonreactive with oxygen.

Physical Vs. Chemical Properties

You can describe matter by both physical and chemical properties. The properties that are most useful in identifying a substance, such as density, solubility, and reactivity with acids, are its characteristic properties. The *characteristic properties* of a substance are always the same whether the sample you're observing is large or small. Scientists rely on characteristic properties to identify and classify substances. **Figure 14** describes some physical and chemical properties.

It is important to remember the differences between physical and chemical properties. For example, you can observe physical properties without changing the identity of the substance. You can observe chemical properties only in situations in which the identity of the substance could change.

Figure 14 *Substances have different physical and chemical properties.*

(a) Helium is used in airships because it is less dense than air and is nonflammable.

(b) If you add bleach to water that is mixed with red food coloring, the red color will disappear.

Comparing Physical and Chemical Properties		
Substance	**Physical property**	**Chemical property**
Helium	less dense than air	nonflammable
Wood	grainy texture	flammable
Baking soda	white powder	reacts with vinegar to produce bubbles
Powdered sugar	white powder	does not react with vinegar
Rubbing alcohol	clear liquid	flammable
Red food coloring	red color	reacts with bleach and loses color
Iron	malleable	reacts with oxygen

Physical Changes Don't Form New Substances

A **physical change** is a change that affects one or more physical properties of a substance. For example, if you break a piece of chalk in two, you change its physical properties of size and shape. But no matter how many times you break it, chalk is still chalk. The chemical properties of the chalk remain unchanged. Each piece of chalk would still produce bubbles if you placed it in vinegar.

Examples of Physical Changes Melting is a good example of a physical change, as you can see in **Figure 15.** Still another physical change occurs when a substance dissolves into another substance. If you dissolve sugar in water, the sugar seems to disappear into the water. But the identity of the sugar does not change. If you taste the water, you will also still taste the sugar. The sugar has undergone a physical change. See the chart below for more examples of physical changes.

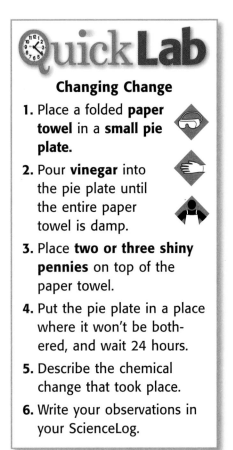

Figure 15 *A physical change turned a stick of butter into the liquid butter that makes popcorn so tasty, but the identity of the butter did not change.*

More Examples of Physical Changes

- Freezing water for ice cubes
- Sanding a piece of wood
- Cutting your hair
- Crushing an aluminum can
- Bending a paper clip
- Mixing oil and vinegar

Can Physical Changes Be Undone? Because physical changes do not change the identity of substances, they are often easy to undo. If you leave butter out on a warm counter, it will undergo a physical change—it will melt. Putting it back in the refrigerator will reverse this change. Likewise, if you create a figure from a lump of clay, you change the clay's shape, causing a physical change. But because the identity of the clay does not change, you can crush your creation and form the clay back into its previous shape.

Chemical Changes Form New Substances

A **chemical change** occurs when one or more substances are changed into entirely new substances with different properties. Chemical changes will or will not occur as described by the chemical properties of substances. But chemical changes and chemical properties are not the same thing. A chemical property describes a substance's ability to go through a chemical change; a chemical change is the actual process in which that substance changes into another substance. You can observe chemical properties only when a chemical change might occur.

QuickLab

Changing Change

1. Place a folded **paper towel** in a **small pie plate.**
2. Pour **vinegar** into the pie plate until the entire paper towel is damp.
3. Place **two or three shiny pennies** on top of the paper towel.
4. Put the pie plate in a place where it won't be bothered, and wait 24 hours.
5. Describe the chemical change that took place.
6. Write your observations in your ScienceLog.

A fun (and delicious) way to see what happens during chemical changes is to bake a cake. When you bake a cake, you combine eggs, flour, sugar, butter, and other ingredients as shown in **Figure 16.** Each ingredient has its own set of properties. But if you mix them together and bake the batter in the oven, you get something completely different. The heat of the oven and the interaction of the ingredients cause a chemical change. As shown in **Figure 17,** you get a cake that has properties completely different to any of the ingredients. Some more examples of chemical changes are shown below.

Figure 16 *Each of these ingredients has different physical and chemical properties.*

Figure 17 *Chemical changes produce new substances with different properties.*

Examples of Chemical Changes

Soured milk smells bad because bacteria have formed new substances in the milk.

Effervescent tablets bubble when the citric acid and baking soda in them react with water.

The hot gas formed when hydrogen and oxygen join to make water helps blast the space shuttle into orbit.

The Statue of Liberty is made of shiny, orange-brown copper. But the metal's interaction with carbon dioxide and water has formed a new substance, copper carbonate, and made this landmark lady green over time.

Clues to Chemical Changes Look back at the bottom of the previous page. In each picture, there is at least one clue that signals a chemical change. Can you find the clues? Here's a hint: chemical changes often cause color changes, fizzing or foaming, heat, or the production of sound, light, or odor.

In the cake example, you would probably smell the sweet aroma of the cake as it baked. If you looked into the oven, you would see the batter rise and turn brown. When you cut the finished cake, you would see the spongy texture created by gas bubbles that formed in the batter (if you baked it right, that is!). All of these yummy clues are signals of chemical changes. But are the clues and the chemical changes the same thing? No, the clues just result from the chemical changes.

Can Chemical Changes Be Undone? Because new substances are formed, you cannot reverse chemical changes using physical means. In other words, you can't uncrumple or iron out a chemical change. Imagine trying to un-bake the cake shown in **Figure 18** by pulling out each ingredient. No way! Most of the chemical changes you see in your daily life, such as a cake baking or milk turning sour, would be difficult to reverse. However, some chemical changes can be reversed under the right conditions by other chemical changes. For example, the water formed in the space shuttle's rockets could be split back into hydrogen and oxygen using an electric current.

Environment

C O N N E C T I O N

When fossil fuels are burned, a chemical change takes place involving sulfur (a substance in fossil fuels) and oxygen (from the air). This chemical change produces sulfur dioxide, a gas. When sulfur dioxide enters the atmosphere, it undergoes another chemical change by interacting with water and oxygen. This chemical change produces sulfuric acid, a contributor to acid precipitation. Acid precipitation can kill trees and make ponds and lakes unable to support life.

Figure 18 *Looking for the original ingredients? You won't find them—their identities have changed.*

SECTION REVIEW

1. Classify each of the following properties as either physical or chemical: reacts with water, dissolves in acetone, is blue, does not react with hydrogen.

2. List three clues that indicate a chemical change might be taking place.

3. **Comparing Concepts** Describe the difference between physical changes and chemical changes in terms of what happens to the matter involved in each kind of change.

internetconnect

SC*i*LINKS.
NSTA

TOPIC: Describing Matter
GO TO: www.scilinks.org
*sci*LINKS **NUMBER:** HSTP035

White Before Your Eyes

You have learned how to describe matter based on its physical and chemical properties. You also have learned some clues that can help you determine whether a change in matter is a physical change or a chemical change. In this lab, you'll use what you have learned to describe four substances, based on their properties and the changes they undergo.

MATERIALS

- 4 spatulas
- baking powder
- plastic-foam egg carton
- 3 eyedroppers
- water
- stirring rod
- vinegar
- iodine solution
- baking soda
- cornstarch
- sugar

Procedure

1. Copy Tables 1 and 2, shown on the next page, into your ScienceLog. Be sure to leave plenty of room in each box to write down your observations. Before you start the lab, put on your safety goggles.

2. Use a spatula to place a small amount of baking powder (just enough to cover the bottom of the cup) into three cups of your egg carton. Look closely at the baking powder. Record your observations about its appearance, such as color and texture, in Table 1 in the column titled "Unmixed."

3. Use an eyedropper to add 60 drops of water to the baking powder in the first cup. Stir the mixture with the stirring rod. Record your observations in Table 1 in the column titled "Mixed with water." Clean your stirring rod.

4. Use a clean dropper to add 20 drops of vinegar to the second cup of baking powder. Stir. Record your observations in Table 1 in the column titled "Mixed with vinegar." Clean your stirring rod.

5. Use a clean dropper to add 5 drops of iodine solution to the third cup of baking powder. Stir. Record your observations in Table 1 in the column titled "Mixed with iodine solution." Clean your stirring rod.
   Caution: Be careful when using iodine. Iodine will stain your skin and clothes.

6. Repeat steps 2–5 for each of the other substances (baking soda, cornstarch, and sugar). Use a clean spatula for each substance.

Table 1 Observations				
Substance	Unmixed	Mixed with water	Mixed with vinegar	Mixed with iodine solution
Baking powder				
Baking soda				
Cornstarch				
Sugar				

DO NOT WRITE IN BOOK

Table 2 Changes and Properties						
	Mixed with water		Mixed with vinegar		Mixed with iodine solution	
Substance	Change	Property	Change	Property	Change	Property
Baking powder						
Baking soda						
Cornstarch						
Sugar						

DO NOT WRITE IN BOOK

Analysis

7 What physical properties do all four substances share?

8 In Table 2, write the type of change you observed, and state the property that the change demonstrates.

9 Classify the four substances you tested by their chemical properties. For example, which substances are reactive with vinegar (acid)?

Chapter Highlights

Vocabulary

matter *(p. 4)*

volume *(p. 4)*

meniscus *(p. 5)*

mass *(p. 6)*

gravity *(p. 7)*

weight *(p. 8)*

newton *(p. 9)*

inertia *(p. 10)*

Section Notes

• Matter is anything that has volume and mass.

• Volume is the amount of space taken up by an object.

• The volume of liquids is expressed in liters and milliliters.

• The volume of solid objects is expressed in cubic units, such as cubic meters.

• Mass is the amount of matter that something is made of.

• Mass and weight are not the same thing. Weight is a measure of the gravitational force exerted on an object, usually in relation to the Earth.

• Mass is usually expressed in milligrams, grams, and kilograms.

• The newton is the SI unit of force, so weight is expressed in newtons.

• Inertia is the tendency of all objects to resist any change in motion. Mass is a measure of inertia. The more massive an object is, the greater its inertia.

Labs

Volumania! *(p. 134)*

Measuring Liquid Volume *(p. 132)*

Coin Operated *(p. 133)*

☑ Skills Check

Math Concepts

DENSITY To calculate an object's density, divide the mass of the object by its volume. For example, the density of an object with a mass of 45 g and a volume of 5.5 cm³ is calculated as follows:

$$D = \frac{m}{V}$$
$$D = \frac{45 \text{ g}}{5.5 \text{ cm}^3}$$
$$D = 8.2 \text{ g/cm}^3$$

Visual Understanding

MASS AND WEIGHT
Mass and weight are related, but they're not the same thing. Look back at Figure 8 on page 9 to learn about the differences between mass and weight.

PHYSICAL AND CHEMICAL PROPERTIES All substances have physical and chemical properties. You can compare some of those properties by reviewing the table on page 16.

Vocabulary

physical property *(p. 11)*
density *(p. 12)*
chemical property *(p. 15)*
physical change *(p. 16)*
chemical change *(p. 17)*

Section Notes

- Physical properties of matter can be observed without changing the identity of the matter.

- Density is the amount of matter in a given space, or the mass per unit volume.

- The density of a substance is always the same at a given pressure and temperature regardless of the size of the sample of the substance.

- Chemical properties describe a substance based on its ability to change into a new substance with different properties.

- Chemical properties can be observed only when one substance might become a new substance.

- The characteristic properties of a substance are always the same whether the sample observed is large or small.

- When a substance undergoes a physical change, its identity remains the same.

- A chemical change occurs when one or more substances are changed into new substances with different properties.

Labs

Determining Density *(p. 136)*
Layering Liquids *(p. 137)*

---

internet connect

GO TO: go.hrw.com

Visit the **HRW** Web site for a variety of learning tools related to this chapter. Just type in the keyword:

KEYWORD: HSTMAT

GO TO: www.scilinks.org

Visit the **National Science Teachers Association** on-line Web site for Internet resources related to this chapter. Just type in the *sci*LINKS number for more information about the topic:

TOPIC: What Is Matter?   *sci*LINKS NUMBER: HSTP030
TOPIC: Describing Matter   *sci*LINKS NUMBER: HSTP035
TOPIC: Dark Matter   *sci*LINKS NUMBER: HSTP040
TOPIC: Building a Better Body   *sci*LINKS NUMBER: HSTP045

Chapter Review

USING VOCABULARY

For each pair of terms, explain the difference in their meanings.

1. mass/volume

2. mass/weight

3. inertia/mass

4. volume/density

5. physical property/chemical property

6. physical change/chemical change

UNDERSTANDING CONCEPTS

Multiple Choice

7. Which of these is *not* matter?
   a. a cloud   c. sunshine
   b. your hair   d. the sun

8. The mass of an elephant on the moon would be
   a. less than its mass on Mars.
   b. more than its mass on Mars.
   c. the same as its weight on the moon.
   d. None of the above

9. Which of the following is *not* a chemical property?
   a. reactivity with oxygen
   b. malleability
   c. flammability
   d. reactivity with acid

10. Your weight could be expressed in which of the following units?
    a. pounds
    b. newtons
    c. kilograms
    d. both (a) and (b)

11. You accidentally break your pencil in half. This is an example of
    a. a physical change.
    b. a chemical change.
    c. density.
    d. volume.

12. Which of the following statements about density is true?
    a. Density depends on mass and volume.
    b. Density is weight per unit volume.
    c. Density is measured in milliliters.
    d. Density is a chemical property.

13. Which of the following pairs of objects would have the greatest attraction toward each other due to gravity?
    a. a 10 kg object and a 10 kg object, 4 m apart
    b. a 5 kg object and a 5 kg object, 4 m apart
    c. a 10 kg object and a 10 kg object, 2 m apart
    d. a 5 kg object and a 5 kg object, 2 m apart

14. Inertia increases as ___?___ increases.
    a. time   c. mass
    b. length   d. volume

Short Answer

15. In one or two sentences, explain the different processes in measuring the volume of a liquid and measuring the volume of a solid.

16. In one or two sentences, explain the relationship between mass and inertia.

17. What is the formula for calculating density?

18. List three characteristic properties of matter.

Concept Mapping

19. Use the following terms to create a concept map: matter, mass, inertia, volume, milliliters, cubic centimeters, weight, gravity.

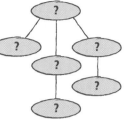

CRITICAL THINKING AND PROBLEM SOLVING

20. You are making breakfast for your picky friend, Filbert. You make him scrambled eggs. He asks, "Would you please take these eggs back to the kitchen and poach them?" What scientific reason do you give Filbert for not changing his eggs?

Poach these, please!

21. You look out your bedroom window and see your new neighbors moving in. Your neighbor bends over to pick up a small cardboard box, but he cannot lift it. What can you conclude about the item(s) in the box? Use the terms *mass* and *inertia* to explain how you came to this conclusion.

22. You may sometimes hear on the radio or on television that astronauts are "weightless" in space. Explain why this is not true.

23. People commonly use the term *volume* to describe the capacity of a container. How does this definition of volume differ from the scientific definition?

MATH IN SCIENCE

24. What is the volume of a book with the following dimensions: a width of 10 cm, a length that is two times the width, and a height that is half the width? Remember to express your answer in cubic units.

25. A jar contains 30 mL of glycerin (mass = 37.8 g) and 60 mL of corn syrup (mass = 82.8 g). Which liquid is on top? Show your work, and explain your answer.

INTERPRETING GRAPHICS

Examine the photograph below, and answer the following questions:

26. List three physical properties of this can.

27. Did a chemical change or a physical change cause the change in this can's appearance?

28. How does the density of the metal in the can compare before and after the change?

29. Can you tell what the chemical properties of the can are just by looking at the picture? Explain.

Reading Check-up

Take a minute to review your answers to the Pre-Reading Questions found at the bottom of page 2. Have your answers changed? If necessary, revise your answers based on what you have learned since you began this chapter.

In the Dark About Dark Matter

What is the universe made of? Believe it or not, when astronomers try to answer this question, they still find themselves in the dark. Surprisingly, there is more to the universe than meets the eye.

A Matter of Gravity

Astronomers noticed something odd when studying the motions of galaxies in space. They expected to find a lot of mass in the galaxies. Instead, they discovered that the mass of the galaxies was not great enough to explain the large gravitational force causing the galaxies' rapid rotation. So what was causing the additional gravitational force? Some scientists think the universe contains matter that we cannot see with our eyes or our telescopes. Astronomers call this invisible matter *dark matter.*

Dark matter doesn't reveal itself by giving off any kind of electromagnetic radiation, such as visible light, radio waves, or gamma radiation. According to scientific calculations, dark matter could account for between 90 and 99 percent of the total mass of the universe! What is dark matter? Would you believe MACHOs and WIMPs?

MACHOs

Scientists recently proved the existence of *MAssive Compact Halo Objects* (MACHOs) in our Milky Way galaxy by measuring their gravitational effects. Even though scientists know MACHOs exist, they aren't sure what MACHOs are made of. Scientists suggest that MACHOs may be brown dwarfs, old white dwarfs, neutron stars, or black holes. Others suggest they

▲ *The Large Magellanic Cloud, located 180,000 light-years from Earth*

are some type of strange, new object whose properties still remain unknown. Even though the number of MACHOs is apparently very great, they still do not represent enough missing mass. So scientists offer another candidate for dark matter—WIMPs.

WIMPs

Theories predict that *Weakly Interacting Massive Particles* (WIMPs) exist, but scientists have never detected them. WIMPs are thought to be massive elementary particles that do not interact strongly with normal matter (which is why scientists have not found them).

More Answers Needed

So far, evidence supports the existence of MACHOs, but there is little or no solid evidence of WIMPs or any other form of dark matter. Scientists who support the idea of WIMPs are conducting studies of the particles that make up matter to see if they can detect WIMPs. Other theories are that gravity acts differently around galaxies or that the universe is filled with things called "cosmic strings." Scientists admit they have a lot of work to do before they will be able to describe the universe—and all the matter in it.

On Your Own

▶ What is microlensing, and what does it have to do with MACHOs? How might the neutrino provide valuable information to scientists who are interested in proving the existence of WIMPs? Find out on your own!

Building a Better Body

Have you ever broken an arm or a leg? If so, you probably wore a cast while the bone healed. But what happens when a bone is too badly damaged to heal? In some cases, a false bone made from a metal called titanium can take the original bone's place. Could using titanium bone implants be the first step in creating bionic body parts? Think about it as you read about some of titanium's amazing properties.

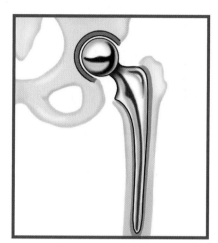

▲ *Titanium bones—even better than the real thing?*

Imitating the Original

Why would a metal like titanium be used to imitate natural bone? Well, it turns out that a titanium implant passes some key tests for bone replacement. First of all, real bones are incredibly lightweight and sturdy, and healthy bones last for many years. Therefore, a bone-replacement material has to be lightweight but also very durable. Titanium passes this test because it is well known for its strength, and it is also lightweight.

Second, the human body's immune system is always on the lookout for foreign substances. If a doctor puts a false bone in place and the patient's immune system attacks it, an infection can result. Somehow, the false bone must be able to chemically trick the body into thinking that the bone is real. Does titanium pass this test? Keep reading!

Accepting Imitation

By studying the human body's immune system, scientists found that the body accepts certain metals. The body almost always accepts one metal in particular. Yep, you guessed it—titanium! This turned out to be quite a discovery.

Doctors could implant pieces of titanium into a person's body without triggering an immune reaction. A bond can even form between titanium and existing bone tissue, fusing the bone to the metal!

Titanium is shaping up to be a great bone-replacement material. It is lightweight and strong, is accepted by the body, can attach to existing bone, and resists chemical changes, such as corrosion. But scientists have encountered a slight problem. Friction can wear away titanium bones, especially those used near the hips and elbows.

Real Success

An unexpected surprise, not from the field of medicine but from the field of nuclear physics, may have solved the problem. Researchers have learned that by implanting a special form of nitrogen on the surface of a piece of metal, they can create a surface layer on the metal that is especially durable and wear-resistant. When this form of nitrogen is implanted in titanium bones, the bones retain all the properties of pure titanium bones but also become very wear-resistant. The new bones should last through decades of heavy use without needing to be replaced.

Think About It

▶ What will the future hold? As time goes by, doctors become more successful at implanting titanium bones. What do you think would happen if the titanium bones were to eventually become better than real bones?

States of Matter

Pre-Reading Questions

1. What are the four most familiar states of matter?

2. Compare the motion of particles in a solid, a liquid, and a gas.

3. Name three ways matter changes from one state to another.

IT TAKES METTLE TO MELT METAL

If you wanted to make a flavored ice pop, you would pour juice into a mold and freeze it. You are able to make the ice pop into the desired shape because, unlike solids, liquids will take the shape of their container. Metal workers apply this important property of liquids when they create metal parts that have complicated shapes. They melt the metal at extremely high temperatures and then pour it into a mold. In this chapter, you will find out more about the properties of different states of matter.

VANISHING ACT

In this activity, you will use rubbing alcohol to investigate a change of state.

Procedure

1. Pour **rubbing alcohol** into a **small plastic cup** until the alcohol just covers the bottom of the cup.

2. Moisten the tip of a **cotton swab** by dipping it into the alcohol in the cup.

3. Rub the cotton swab on the palm of your hand.

4. Record your observations in your ScienceLog.

5. Wash your hands thoroughly.

Analysis

6. Explain what happened to the alcohol.

7. Did you feel a sensation of hot or cold? If so, how do you explain what you observed?

8. Record your answers in your ScienceLog.

Terms to Learn

states of matter   pressure
solid   Boyle's law
liquid   Charles's law
gas   plasma

What You'll Do

- Describe the properties shared by particles of all matter.
- Describe the four states of matter discussed here.
- Describe the differences between the states of matter.
- Predict how a change in pressure or temperature will affect the volume of a gas.

Four States of Matter

Figure 1 shows a model of the earliest known steam engine, invented about A.D. 60 by Hero, a scientist who lived in Alexandria, Egypt. This model also demonstrates the four most familiar states of matter: solid, liquid, gas, and plasma. The **states of matter** are the physical forms in which a substance can exist. For example, water commonly exists in three different states of matter: solid (ice), liquid (water), and gas (steam).

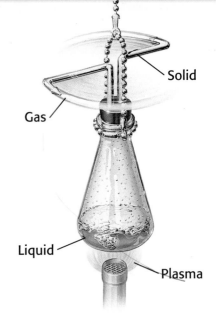

Figure 1 *This model of Hero's steam engine spins as steam escapes through the nozzles.*

Moving Particles Make Up All Matter

Matter consists of tiny particles called atoms and molecules (MAHL i KYOOLZ) that are too small to see without an amazingly powerful microscope. These atoms and molecules are always in motion and are constantly bumping into one another. The state of matter of a substance is determined by how fast the particles move and how strongly the particles are attracted to one another. **Figure 2** illustrates three of the states of matter—solid, liquid, and gas—in terms of the speed and attraction of the particles.

Figure 2 Models of a Solid, a Liquid, and a Gas

Particles of a solid do not move fast enough to overcome the strong attraction between them, so they are held tightly in place. The particles vibrate in place.

Particles of a liquid move fast enough to overcome some of the attraction between them. The particles are able to slide past one another.

Particles of a gas move fast enough to overcome nearly all of the attraction between them. The particles move independently of one another.

Solids Have Definite Shape and Volume

Look at the ship in **Figure 3.** Even in a bottle, it keeps its original shape and volume. If you moved the ship to a larger bottle, the ship's shape and volume would not change. Scientifically, the state in which matter has a definite shape and volume is **solid.** The particles of a substance in a solid are very close together. The attraction between them is stronger than the attraction between the particles of the same substance in the liquid or gaseous state. The atoms or molecules in a solid move, but not fast enough to overcome the attraction between them. Each particle vibrates in place because it is locked in position by the particles around it.

Figure 3 *Because this ship is a solid, it does not take the shape of the bottle.*

Two Types of Solids Solids are often divided into two categories—*crystalline* and *amorphous* (uh MOHR fuhs). Crystalline solids have a very orderly, three-dimensional arrangement of atoms or molecules. That is, the particles are arranged in a repeating pattern of rows. Examples of crystalline solids include iron, diamond, and ice. Amorphous solids are composed of atoms or molecules that are in no particular order. That is, each particle is in a particular spot, but the particles are in no organized pattern. Examples of amorphous solids include rubber and wax. **Figure 4** illustrates the differences in the arrangement of particles in these two solids.

Activity

Imagine that you are a particle in a solid. Your position in the solid is your chair. In your ScienceLog, describe the different types of motion that are possible even though you cannot leave your chair.

Figure 4 *Differing arrangements of particles in crystalline solids and amorphous solids lead to different properties. Imagine trying to hit a home run with a rubber bat!*

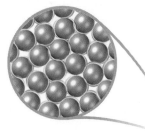

The particles in a **crystalline solid** have a very orderly arrangement.

The particles in an **amorphous solid** do not have an orderly arrangement.

Liquids Change Shape but Not Volume

A liquid will take the shape of whatever container it is put in. You are reminded of this every time you pour yourself a glass of juice. The state in which matter takes the shape of its container and has a definite volume is **liquid.** The atoms or molecules in liquids move fast enough to overcome some of the attractions between them. The particles slide past each other until the liquid takes the shape of its container. **Figure 5** shows how the particles in juice might look if they were large enough to see.

Even though liquids change shape, they do not readily change volume. You know that a can of soda contains a certain volume of liquid regardless of whether you pour it into a large container or a small one. **Figure 6** illustrates this point using a beaker and a graduated cylinder.

Figure 5 *Particles in a liquid slide past one another until the liquid conforms to the shape of its container.*

Figure 6 *Even when liquids change shape, they don't change volume.*

The Squeeze Is On Because the particles in liquids are close to one another, it is difficult to push them closer together. This makes liquids ideal for use in hydraulic (hie DRAW lik) systems. For example, brake fluid is the liquid used in the brake systems of cars. Stepping on the brake pedal applies a force to the liquid. The particles in the liquid move away rather than squeezing closer together. As a result, the fluid pushes the brake pads outward against the wheels, which slows the car.

BRAIN FOOD

The Boeing 767 Freighter, a type of commercial airliner, has 187 km (116 mi) of hydraulic tubing.

A Drop in the Bucket Two other important properties of liquids are *surface tension* and *viscosity* (vis KAHS uh tee). Surface tension is the force acting on the particles at the surface of a liquid that causes the liquid to form spherical drops, as shown in **Figure 7.** Different liquids have different surface tensions. For example, rubbing alcohol has a lower surface tension than water, but mercury has a higher surface tension than water.

Viscosity is a liquid's resistance to flow. In general, the stronger the attractions between a liquid's particles are, the more viscous the liquid is. Think of the difference between pouring honey and pouring water. Honey flows more slowly than water because it has a higher viscosity than water.

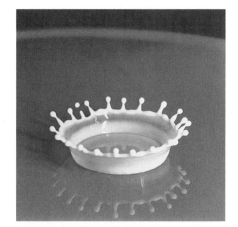

Figure 7 *Liquids form spherical drops as a result of surface tension.*

Gases Change Both Shape and Volume

How many balloons can be filled from a single metal cylinder of helium? The number may surprise you. One cylinder can fill approximately 700 balloons. How is this possible? After all, the volume of the metal cylinder is equal to the volume of only about five inflated balloons.

It's a Gas! Helium is a gas. **Gas** is the state in which matter changes in both shape and volume. The atoms or molecules in a gas move fast enough to break away completely from one another. Therefore, the particles of a substance in the gaseous state have less attraction between them than particles of the same substance in the solid or liquid state. In a gas, there is empty space between particles.

The amount of empty space in a gas can change. For example, the helium in the metal cylinder consists of atoms that have been forced very close together, as shown in **Figure 8.** As the helium fills the balloon, the atoms spread out, and the amount of empty space in the gas increases. As you continue reading, you will learn how this empty space is related to pressure.

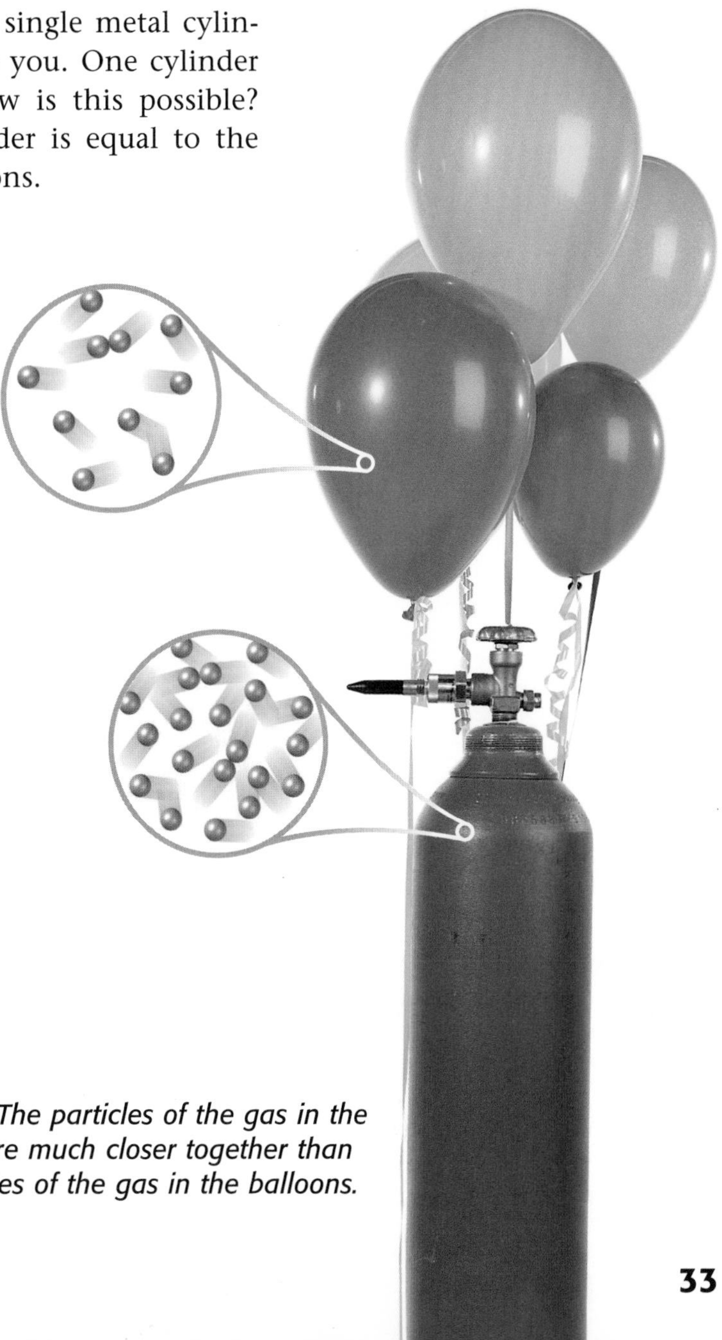

Figure 8 *The particles of the gas in the cylinder are much closer together than the particles of the gas in the balloons.*

33

Gas Under Pressure

Pressure is the amount of force exerted on a given area. You can think of this as the number of collisions of particles against the inside of the container. Compare the basketball with the beach ball in **Figure 9.** The balls have the same volume and contain particles of gas (air) that constantly collide with one another and with the inside surface of the balls. Notice, however, that there are more particles in the basketball than in the beach ball. As a result, more particles collide with the inside surface of the basketball than with the inside surface of the beach ball. When the number of collisions increases, the force on the inside surface of the ball increases. This increased force leads to increased pressure.

✓ Self-Check

How would an increase in the speed of the particles affect the pressure of gas in a metal cylinder? *(See page 168 to check your answer.)*

Figure 9 *Both balls shown here are full of air, but the pressure in the basketball is higher than the pressure in the beach ball.*

The basketball has a higher pressure than the beach ball because the greater number of particles of gas are closer together. Therefore, they collide with the inside of the ball at a faster rate.

The beach ball has a lower pressure than the basketball because the lesser number of particles of gas are farther apart. Therefore, they collide with the inside of the ball at a slower rate.

🖳 internet connect

SCILINKS.
NSTA

TOPIC: Solids, Liquids, and Gases
GO TO: www.scilinks.org
*sci*LINKS NUMBER: HSTP060

SECTION REVIEW

1. List two properties that all particles of matter have in common.

2. Describe solids, liquids, and gases in terms of shape and volume.

3. Why can the volume of a gas change?

4. **Applying Concepts** Explain what happens inside the ball when you pump up a flat basketball.

Laws Describe Gas Behavior

Earlier in this chapter, you learned about the atoms and molecules in both solids and liquids. You learned that compared with gas particles, the particles of solids and liquids are closely packed together. As a result, solids and liquids do not change volume very much. Gases, on the other hand, behave differently; their volume can change by a large amount.

It is easy to measure the volume of a solid or liquid, but how do you measure the volume of a gas? Isn't the volume of a gas the same as the volume of its container? The answer is yes, but there are other factors, such as pressure, to consider.

Boyle's Law Imagine a diver at a depth of 10 m blowing a bubble of air. As the bubble rises, its volume increases. By the time the bubble reaches the surface, its original volume will have doubled due to the decrease in pressure. The relationship between the volume and pressure of a gas is known as Boyle's law because it was first described by Robert Boyle, a seventeenth-century Irish chemist. **Boyle's law** states that for a fixed amount of gas at a constant temperature, the volume of a gas increases as its pressure decreases. Likewise, the volume of a gas decreases as its pressure increases. Boyle's law is illustrated by the model in **Figure 10.**

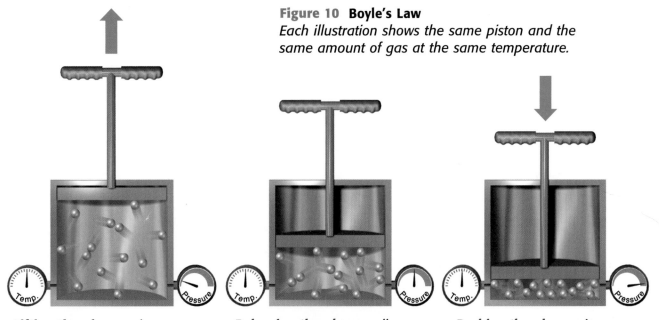

Figure 10 Boyle's Law
Each illustration shows the same piston and the same amount of gas at the same temperature.

Lifting the plunger decreases the pressure of the gas. The particles of gas spread farther apart. The volume of the gas increases as the pressure decreases.

Releasing the plunger allows the gas to change to an intermediate volume and pressure.

Pushing the plunger increases the pressure of the gas. The particles of gas are forced closer together. The volume of the gas decreases as the pressure increases.

See Charles's law in action for yourself using a balloon on page 138 of the LabBook.

Weather balloons demonstrate a practical use of Boyle's law. A weather balloon carries equipment into the atmosphere to collect information used to predict the weather. This balloon is filled with only a small amount of gas because the pressure of the gas decreases and the volume increases as the balloon rises. If the balloon were filled with too much gas, it would pop as the volume of the gas increased.

Charles's Law An inflated balloon will also pop when it gets too hot, demonstrating another gas law—Charles's law. **Charles's law** states that for a fixed amount of gas at a constant pressure, the volume of the gas increases as its temperature increases. Likewise, the volume of the gas decreases as its temperature decreases. Charles's law is illustrated by the model in **Figure 11.** You can see Charles's law in action by putting an inflated balloon in the freezer. Wait about 10 minutes, and see what happens!

MATH BREAK

Gas Law Graphs

Each graph below illustrates a gas law. However, the variable on one axis of each graph is not labeled. Answer the following questions for each graph:

1. As the volume increases, what happens to the missing variable?

2. Which gas law is shown?

3. What label belongs on the axis?

4. Is the graph linear or non-linear? What does this tell you?

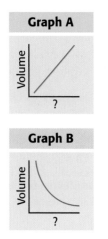

Figure 11 Charles's Law
Each illustration shows the same piston and the same amount of gas at the same pressure.

Lowering the temperature of the gas causes the particles to move more slowly. They hit the sides of the piston less often and with less force. As a result, the volume of the gas decreases.

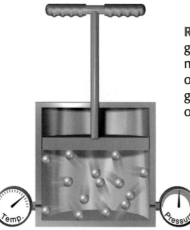

Raising the temperature of the gas causes the particles to move more quickly. They hit the sides of the piston more often and with greater force. As a result, the volume of the gas increases.

Charles's Law and Bicycle Tires

One of your friends overinflated the tires on her bicycle. Use Charles's law to explain why she should let out some of the air before going for a ride on a hot day.

Plasmas

Scientists estimate that more than 99 percent of the known matter in the universe, including the sun and other stars, is made of a state of matter called plasma. **Plasma** is the state of matter that does not have a definite shape or volume and whose particles have broken apart.

Plasmas have some properties that are quite different from the properties of gases. Plasmas conduct electric current, while gases do not. Electric and magnetic fields affect plasmas but do not affect gases. In fact, strong magnetic fields are used to contain very hot plasmas that would destroy any other container.

Natural plasmas are found in lightning, fire, and the incredible light show in **Figure 12,** called the aurora borealis (ah ROHR uh BOHR ee AL is). Artificial plasmas, found in fluorescent lights and plasma balls, are created by passing electric charges through gases.

Figure 12 *Auroras, like the aurora borealis seen here, form when high-energy plasma collides with gas particles in the upper atmosphere.*

SECTION REVIEW

1. When scientists record the volume of a gas, why do they also record the temperature and the pressure?

2. List two differences between gases and plasmas.

3. **Applying Concepts** What happens to the volume of a balloon left on a sunny windowsill? Explain.

internetconnect

SC*i*LINKS
NSTA

TOPIC: Natural and Artificial Plasma
GO TO: www.scilinks.org
*sci*LINKS NUMBER: HSTP065

Changes of State

Terms to Learn

change of state   boiling
melting   evaporation
freezing   condensation
vaporization   sublimation

What You'll Do

◆ Describe how substances change
   from state to state.
◆ Explain the difference between
   an exothermic change and an
   endothermic change.
◆ Compare the changes of state.

A **change of state** is the conversion of a substance from one physical form to another. All changes of state are physical changes. In a physical change, the identity of a substance does not change. In **Figure 13,** the ice, liquid water, and steam are all the same substance—water. In this section, you will learn about the four changes of state illustrated in Figure 13 as well as a fifth change of state called *sublimation* (SUHB li MAY shuhn).

Figure 13 *The terms in the arrows are changes of state. Water commonly goes through the changes of state shown here.*

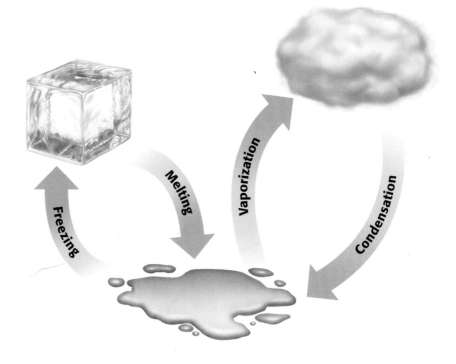

Energy and Changes of State

During a change of state, the energy of a substance changes. The *energy* of a substance is related to the motion of its particles. The molecules in the liquid water in Figure 13 move faster than the molecules in the ice. Therefore, the liquid water has more energy than the ice.

If energy is added to a substance, its particles move faster. If energy is removed, its particles move slower. The *temperature* of a substance is a measure of the speed of its particles and therefore is a measure of its energy. For example, steam has a higher temperature than liquid water, so particles in steam have more energy than particles in liquid water. A transfer of energy, known as *heat,* causes the temperature of a substance to change, which can lead to a change of state.

Want to learn how to get power from changes of state? Steam ahead to page 51.

Melting: Solids to Liquids

Melting is the change of state from a solid to a liquid. This is what happens when an ice cube melts. **Figure 14** shows a metal called gallium melting. What is unusual about this metal is that it melts at around 30°C. Because your normal body temperature is about 37°C, gallium will melt right in your hand!

The *melting point* of a substance is the temperature at which the substance changes from a solid to a liquid. Melting points of substances vary widely. The melting point of gallium is 30°C. Common table salt, however, has a melting point of 801°C.

Most substances have a unique melting point that can be used with other data to identify them. Because the melting point does not change with different amounts of the substance, melting point is a *characteristic property* of a substance.

Figure 14 *Even though gallium is a metal, it would not be very useful as jewelry!*

Absorbing Energy For a solid to melt, particles must overcome some of their attractions to each other. When a solid is at its melting point, any energy it absorbs increases the motion of its atoms or molecules until they overcome the attractions that hold them in place. Melting is an *endothermic* change because energy is absorbed by the substance as it changes state.

Freezing: Liquids to Solids

Freezing is the change of state from a liquid to a solid. The temperature at which a liquid changes into a solid is its *freezing point*. Freezing is the reverse process of melting, so freezing and melting occur at the same temperature, as shown in **Figure 15**.

Removing Energy For a liquid to freeze, the motion of its atoms or molecules must slow to the point where attractions between them overcome their motion. If a liquid is at its freezing point, removing more energy causes the particles to begin locking into place. Freezing is an *exothermic* change because energy is removed from, or taken out of, the substance as it changes state.

Figure 15 *Liquid water freezes at the same temperature that ice melts—0°C.*

If energy is added at 0°C, the ice will melt.

If energy is removed at 0°C, the liquid water will freeze.

Vaporization: Liquids to Gases

One way to experience vaporization (VAY puhr i ZAY shuhn) is to iron a shirt—carefully!—using a steam iron. You will notice steam coming up from the iron as the wrinkles are eliminated. This steam results from the vaporization of liquid water by the iron. **Vaporization** is simply the change of state from a liquid to a gas.

Boiling is vaporization that occurs throughout a liquid. The temperature at which a liquid boils is called its *boiling point*. Like the melting point, the boiling point is a characteristic property of a substance. The boiling point of water is 100°C, whereas the boiling point of liquid mercury is 357°C. **Figure 16** illustrates the process of boiling and a second form of vaporization—evaporation (ee VAP uh RAY shuhn).

Evaporation is vaporization that occurs at the surface of a liquid below its boiling point, as shown in Figure 16. When you perspire, your body is cooled through the process of evaporation. Perspiration is mostly water. Water absorbs energy from your skin as it evaporates. You feel cooler because your body transfers energy to the water. Evaporation also explains why water in a glass on a table disappears after several days.

✔ Self-Check

Is vaporization an endothermic or exothermic change?
(See page 168 to check your answer.)

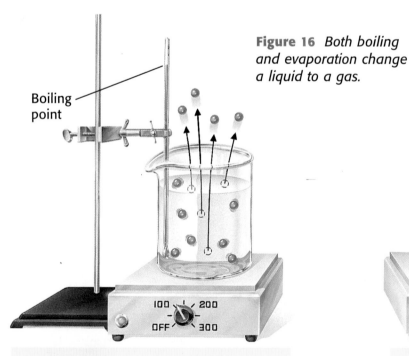

Figure 16 *Both boiling and evaporation change a liquid to a gas.*

Boiling point

Boiling point

Boiling occurs in a liquid at its boiling point. As energy is added to the liquid, particles throughout the liquid move fast enough to break away from the particles around them and become a gas.

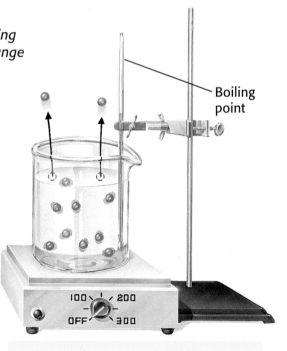

Evaporation occurs in a liquid below its boiling point. Some particles at the surface of the liquid move fast enough to break away from the particles around them and become a gas.

Pressure Affects Boiling Point Earlier you learned that water boils at 100°C. In fact, water only boils at 100°C at sea level because of atmospheric pressure. Atmospheric pressure is caused by the weight of the gases that make up the atmosphere. Atmospheric pressure varies depending on where you are in relation to sea level. Atmospheric pressure is lower at higher elevations. The higher you go above sea level, the less air there is above you, and the lower the atmospheric pressure is. If you were to boil water at the top of a mountain, the boiling point would be lower than 100°C. For example, Denver, Colorado, is 1.6 km (1 mi) above sea level and water boils there at about 95°C. You can make water boil at an even lower temperature by doing the QuickLab at right.

Condensation: Gases to Liquids

Look at the cool glass of lemonade in **Figure 17.** Notice the beads of water on the outside of the glass. These form as a result of condensation. **Condensation** is the change of state from a gas to a liquid. The *condensation point* of a substance is the temperature at which the gas becomes a liquid and is the same temperature as the boiling point at a given pressure. Thus, at sea level, steam condenses to form water at 100°C—the same temperature at which water boils.

For a gas to become a liquid, large numbers of atoms or molecules must clump together. Particles clump together when the attraction between them overcomes their motion. For this to occur, energy must be removed from the gas to slow the particles down. Therefore, condensation is an exothermic change.

Figure 17 *Gaseous water in the air will become liquid when it contacts a cool surface.*

Boiling Water Is Cool

1. Remove the cap from a **syringe.**

2. Place the tip of the syringe in the **warm water** provided by your teacher. Pull the plunger out until you have 10 mL of water in the syringe.

3. Tightly cap the syringe.

4. Hold the syringe, and slowly pull the plunger out.

5. Observe any changes you see in the water. Record your observations in your ScienceLog.

6. Why are you not burned by the boiling water in the syringe?

Meteorology

C O N N E C T I O N

The amount of gaseous water that air can hold decreases as the temperature of the air decreases. As the air cools, some of the gaseous water condenses to form small drops of liquid water. These drops form clouds in the sky and fog near the ground.

Sublimation: Solids Directly to Gases

Look at the solids shown in **Figure 18.** The solid on the left is ice. Notice the drops of liquid collecting as it melts. On the right, you see carbon dioxide in the solid state, also called dry ice. It is called dry ice because instead of melting into a liquid, it goes through a change of state called sublimation. **Sublimation** is the change of state from a solid directly into a gas. Dry ice is colder than ice, and it doesn't melt into a puddle of liquid. It is often used to keep food, medicine, and other materials cold without getting them wet.

For a solid to change directly into a gas, the atoms or molecules must move from being very tightly packed to being very spread apart. The attractions between the particles must be completely overcome. Because this requires the addition of energy, sublimation is an endothermic change.

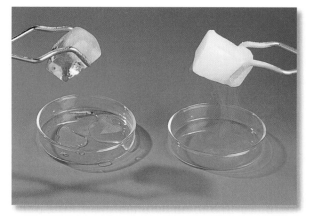

Figure 18 *Ice melts, but dry ice, on the right, turns directly into a gas.*

Comparing Changes of State

As you learned in Section 1 of this chapter, the state of a substance depends on how fast its atoms or molecules move and how strongly they are attracted to each other. A substance may undergo a physical change from one state to another by an endothermic change (if energy is added) or an exothermic change (if energy is removed). The table below shows the differences between the changes of state discussed in this section.

Summarizing the Changes of State			
Change of state	**Direction**	**Endothermic or exothermic?**	**Example**
Melting	solid ⟶ liquid	endothermic	Ice melts into liquid water at 0°C.
Freezing	liquid ⟶ solid	exothermic	Liquid water freezes into ice at 0°C.
Vaporization	liquid ⟶ gas	endothermic	Liquid water vaporizes into steam at 100°C.
Condensation	gas ⟶ liquid	exothermic	Steam condenses into liquid water at 100°C.
Sublimation	solid ⟶ gas	endothermic	Solid dry ice sublimes into a gas at −78°C.

Temperature Change Versus Change of State

When most substances lose or absorb energy, one of two things happens to the substance: its temperature changes or its state changes. Earlier in the chapter, you learned that the temperature of a substance is a measure of the speed of the particles. This means that when the temperature of a substance changes, the speed of the particles also changes. But while a substance changes state, its temperature does not change until the change of state is complete, as shown in **Figure 19.**

Figure 19 Changing the State of Water

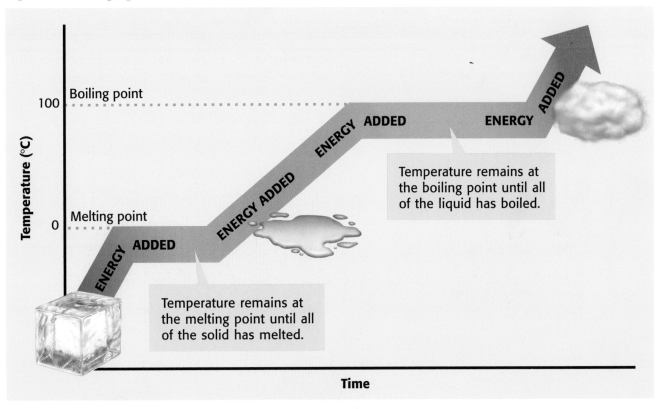

SECTION REVIEW

1. Compare endothermic and exothermic changes.

2. Classify each change of state (melting, freezing, vaporization, condensation, and sublimation) as endothermic or exothermic.

3. Describe how the motion and arrangement of particles change as a substance freezes.

4. **Comparing Concepts**   How are evaporation and boiling different? How are they similar?

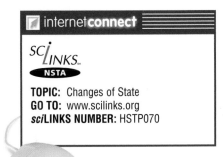

internet**connect**

SC*L*INKS.
NSTA

TOPIC: Changes of State
GO TO: www.scilinks.org
*sci***LINKS NUMBER:** HSTP070

Discovery Lab

A Hot and Cool Lab

When you add energy to a substance through heating, does the substance's temperature always go up? When you remove energy from a substance through cooling, does the substance's temperature always go down? In this lab, you'll investigate these important questions with a very common substance—water.

MATERIALS

- 250 or 400 mL beaker
- water
- heat-resistant gloves
- hot plate
- thermometer
- stopwatch
- 100 mL graduated cylinder
- large coffee can
- crushed ice
- rock salt
- wire-loop stirring device
- graph paper

Form a Hypothesis

1. In your ScienceLog, answer the following questions: What happens to the temperature of boiling water when you continue to add energy through heating? (Part A) What happens to the temperature of freezing water when you continue to remove energy through cooling? (Part B)

Test the Hypothesis (Part A)

2. Make a table like the one on the next page. Label the table "Heating Water."

3. Fill the beaker one-half full with water. Put on heat-resistant gloves. Turn on the hot plate. Put the beaker on the burner. Place the thermometer in the beaker.
   Caution: Do not touch the burner.

Time (s)	30	60	90	120	150	180	210	etc.
Temperature (°C)	DO NOT WRITE IN BOOK							

4 Record the temperature of the water every 30 seconds. Continue taking readings until about one-fourth of the water boils away. Note the first temperature reading at which the water steadily boils.

5 Turn off the hot plate. While the beaker is cooling, make a graph of temperature (*y*-axis) versus time (*x*-axis). Draw an arrow pointing to the first temperature reading at which the water was steadily boiling.

6 After you finish the graph, follow your teacher's instructions for cleanup and disposal.

Test the Hypothesis (Part B)

7 Put 20 mL of water in the graduated cylinder. Place the graduated cylinder in the coffee can, and fill the can with crushed ice. Pour rock salt on the ice. Place the thermometer and the wire-loop stirring device in the graduated cylinder.

8 Label a new table "Cooling Water." Record the temperature of the water in the graduated cylinder every 30 seconds. Add ice and rock salt to the can as needed. Stir the water with the stirring device.
Caution: Do not stir with the thermometer.

9 Once the water begins to freeze, stop stirring. Do not try to pull the thermometer out of the solid ice in the cylinder.

10 Record the temperature when you first see ice crystals forming in the water. Continue taking readings until the water is completely frozen.

11 Make a graph of temperature (*y-axis*) versus time (*x-axis*). Draw an arrow to the temperature reading at which the first ice crystals formed in the water.

Analyze the Results (Parts A and B)

12 What does the slope of each graph represent?

13 How does the slope when the water is boiling compare with the slope before the water boils? Explain why the slopes differ.

14 How does the slope when the water is freezing compare with the slope before the water freezes? Explain why the slopes differ.

Draw Conclusions (Parts A and B)

15 Adding or removing energy leads to changes in the movement of particles that make up solids, liquids, and gases. Use this idea to explain why the temperature graphs of the two experiments look the way they do.

Chapter Highlights

SECTION 1

Vocabulary

states of matter *(p. 30)*

solid *(p. 31)*

liquid *(p. 32)*

gas *(p. 33)*

pressure *(p. 34)*

Boyle's law *(p. 35)*

Charles's law *(p. 36)*

plasma *(p. 37)*

Section Notes

- The states of matter are the physical forms in which a substance can exist. The four most familiar states are solid, liquid, gas, and plasma.

- All matter is made of tiny particles called atoms and molecules that attract each other and move constantly.

- A solid has a definite shape and volume.

- A liquid has a definite volume but not a definite shape.

- A gas does not have a definite shape or volume. A gas takes the shape and volume of its container.

- Pressure is a force per unit area. Gas pressure increases as the number of collisions of gas particles increases.

- Boyle's law states that the volume of a gas increases as the pressure decreases if the temperature does not change.

- Charles's law states that the volume of a gas increases as the temperature increases if the pressure does not change.

- Plasmas are composed of particles that have broken apart. Plasmas do not have a definite shape or volume.

Labs

Full of Hot Air! *(p. 138)*

☑ Skills Check

Math Concepts

GRAPHING DATA The relationship between measured values can be seen by plotting the data on a graph. The top graph shows the linear relationship described by Charles's law—as the temperature of a gas increases, its volume increases. The bottom graph shows the nonlinear relationship described by Boyle's law—as the pressure of a gas increases, its volume decreases.

Visual Understanding

PARTICLE ARRANGEMENT Many of the properties of solids, liquids, and gases are due to the arrangement of the atoms or molecules of the substance. Review the models in Figure 2 on page 30 to study the differences in particle arrangement between the solid, liquid, and gaseous states.

SUMMARY OF THE CHANGES OF STATE Review the table on page 42 to study the direction of each change of state and whether energy is absorbed or removed during each change.

Vocabulary

change of state (*p. 38*)
melting (*p. 39*)
freezing (*p. 39*)
vaporization (*p. 40*)
boiling (*p. 40*)
evaporation (*p. 40*)
condensation (*p. 41*)
sublimation (*p. 42*)

Section Notes

- A change of state is the conversion of a substance from one physical form to another. All changes of state are physical changes.

- Exothermic changes release energy. Endothermic changes absorb energy.

- Melting changes a solid to a liquid. Freezing changes a liquid to a solid. The freezing point and melting point of a substance are the same temperature.

- Vaporization changes a liquid to a gas. There are two kinds of vaporization: boiling and evaporation.

- Boiling occurs throughout a liquid at the boiling point.

- Evaporation occurs at the surface of a liquid, at a temperature below the boiling point.

- Condensation changes a gas to a liquid.

- Sublimation changes a solid directly to a gas.

- Temperature does not change during a change of state.

Labs

Can Crusher (*p. 139*)

---

 internet connect

GO TO: go.hrw.com

Visit the **HRW** Web site for a variety of learning tools related to this chapter. Just type in the keyword:

KEYWORD: HSTSTA

 SCi*LINKS*.

N S T A **GO TO:** www.scilinks.org

Visit the **National Science Teachers Association** on-line Web site for Internet resources related to this chapter. Just type in the *sci*LINKS number for more information about the topic:

TOPIC: Forms and Uses of Glass	*sci***LINKS NUMBER:** HSTP055
TOPIC: Solids, Liquids, and Gases	*sci***LINKS NUMBER:** HSTP060
TOPIC: Natural and Artificial Plasma	*sci***LINKS NUMBER:** HSTP065
TOPIC: Changes of State	*sci***LINKS NUMBER:** HSTP070
TOPIC: The Steam Engine	*sci***LINKS NUMBER:** HSTP075

Chapter Review

For each pair of terms, explain the difference in meaning.

1. solid/liquid

2. Boyle's law/Charles's law

3. evaporation/boiling

4. melting/freezing

UNDERSTANDING CONCEPTS

Multiple Choice

5. Which of the following best describes the particles of a liquid?
   a. The particles are far apart and moving fast.
   b. The particles are close together but moving past each other.
   c. The particles are far apart and moving slowly.
   d. The particles are closely packed and vibrate in place.

6. Boiling points and freezing points are examples of
   a. chemical properties. c. energy.
   b. physical properties.   d. matter.

7. During which change of state do atoms or molecules become more ordered?
   a. boiling   c. melting
   b. condensation   d. sublimation

8. Which of the following describes what happens as the temperature of a gas in a balloon increases?
   a. The speed of the particles decreases.
   b. The volume of the gas increases and the speed of the particles increases.
   c. The volume decreases.
   d. The pressure decreases.

9. Dew collects on a spider web in the early morning. This is an example of
   a. condensation.   c. sublimation.
   b. evaporation.   d. melting.

10. Which of the following changes of state is exothermic?
    a. evaporation   c. freezing
    b. sublimation   d. melting

11. What happens to the volume of a gas inside a piston if the temperature does not change but the pressure is reduced?
    a. increases
    b. stays the same
    c. decreases
    d. not enough information

12. The atoms and molecules in matter
    a. are attracted to one another.
    b. are constantly moving.
    c. move faster at higher temperatures.
    d. All of the above

13. Which of the following contains plasma?
    a. dry ice   c. a fire
    b. steam   d. a hot iron

Short Answer

14. Explain why liquid water takes the shape of its container but an ice cube does not.

15. Rank solids, liquids, and gases in order of decreasing particle speed.

16. Compare the density of iron in the solid, liquid, and gaseous states.

Concept Mapping

17. Use the following terms to create a concept map: states of matter, solid, liquid, gas, plasma, changes of state, freezing, vaporization, condensation, melting.

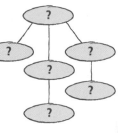

CRITICAL THINKING AND PROBLEM SOLVING

18. After taking a shower, you notice that small droplets of water cover the mirror. Explain how this happens. Be sure to describe where the water comes from and the changes it goes through.

19. In the photo below, water is being split to form two new substances, hydrogen and oxygen. Is this a change of state? Explain your answer.

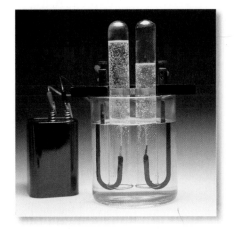

20. To protect their crops during freezing temperatures, orange growers spray water onto the trees and allow it to freeze. In terms of energy lost and energy gained, explain why this practice protects the oranges from damage.

21. At sea level, water boils at 100°C, while methane boils at –161°C. Which of these substances has a stronger force of attraction between its particles? Explain your reasoning.

MATH IN SCIENCE

22. Kate placed 100 mL of water in five different pans, placed the pans on a windowsill for a week, and measured how much water evaporated. Draw a graph of her data, shown below, with surface area on the x-axis. Is the graph linear or nonlinear? What does this tell you?

Pan number	1	2	3	4	5
Surface area (cm²)	44	82	20	30	65
Volume evaporated (mL)	42	79	19	29	62

23. Examine the graph below, and answer the following questions:
    a. What is the boiling point of the substance? What is the melting point?
    b. Which state is present at 30°C?
    c. How will the substance change if energy is added to the liquid at 20°C?

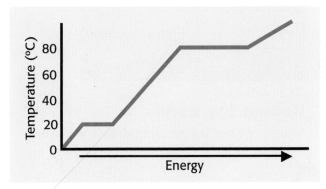

Reading Check-up

Take a minute to review your answers to the Pre-Reading Questions found at the bottom of page 28. Have your answers changed? If necessary, revise your answers based on what you have learned since you began this chapter.

Science, Technology, and Society

Guiding Lightning

By the time you finish reading this sentence, lightning will have flashed more than 500 times around the world. This common phenomenon can have devastating results. Each year in the United States alone, lightning kills almost a hundred people and causes several hundred million dollars in damage. While controlling this awesome outburst of Mother Nature may seem impossible, scientists around the world are searching for ways to reduce the destruction caused by lightning.

Behind the Bolts

Scientists have learned that during a normal lightning strike several events occur. First, electric charges build up at the bottom of a cloud. The cloud then emits a line of negatively charged air particles that zigzags toward the Earth. The attraction between these negatively charged air particles and positively charged particles from objects on the ground forms a *plasma channel.* This channel is the pathway for a lightning bolt. As soon as the plasma channel is complete, BLAM!—between 3 and 20 lightning bolts separated by thousandths of a second travel along it.

A Stroke of Genius

Armed with this information, scientists have begun thinking of ways to redirect these naturally occurring plasma channels. One idea is to use laser beams. In theory, a laser beam directed into a thundercloud can charge the air particles in its path, causing a plasma channel to develop and forcing lightning to strike.

By creating the plasma channels themselves, scientists can, in a way, catch a bolt of lightning before it strikes and direct it to a safe area of the ground. So scientists simply use lasers to direct naturally occurring lightning to strike where they want it to.

A Bright Future?

Laser technology is not without its problems, however. The machines that generate laser beams are large and expensive, and they can themselves be struck by misguided lightning bolts. Also, it is not clear whether creating these plasma channels will be enough to prevent the devastating effects of lightning.

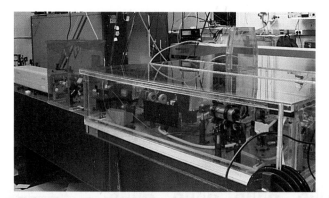

▲ *Sometime in the future, a laser like this might be used to guide lightning away from sensitive areas.*

Find Out for Yourself

▶ Use the Internet or an electronic database to find out how rockets have been used in lightning research. Share your findings with the class.

Eureka!

Full Steam Ahead!

It was huge. It was 40 m long, about 5 m high, and it weighed 245 metric tons. It could pull a 3.28 million kilogram train at 100 km/h. It was a 4-8-8-4 locomotive, called a Big Boy, delivered in 1941 to the Union Pacific Railroad in Omaha, Nebraska. It was also one of the final steps in a 2,000-year search to harness steam power.

A Simple Observation

For thousands of years, people used wind, water, gravity, dogs, horses, and cattle to replace manual labor. But until about 300 years ago, they had limited success. Then in 1690, Denis Papin, a French mathematician and physicist, observed that steam expanding in a cylinder pushed a piston up. As the steam then cooled and contracted, the piston fell. Watching the motion of the piston, Papin had an idea: attach a water-pump handle to the piston. As the pump handle rose and fell with the piston, water was pumped.

More Uplifting Ideas

Eight years later, an English naval captain named Thomas Savery made Papin's device more efficient by using water to cool and condense the steam. Savery's improved pump was used in British coal mines. As good as Savery's pump was, the development of steam power didn't stop there!

In 1712, an English blacksmith named Thomas Newcomen improved Savery's device by adding a second piston and a horizontal beam that acted like a seesaw. One end of the beam was attached to the piston in the steam cylinder. The other end of the beam was attached to the pump piston. As the steam piston moved up and down, it created a vacuum in the pump cylinder and sucked water up from the mine. Newcomen's engine was the most widely used steam engine for more than 50 years.

Watt a Great Idea!

In 1764, James Watt, a Scottish technician, was repairing a Newcomen engine. He realized that heating the cylinder, letting it cool, then heating it again wasted an enormous amount of energy. Watt added a separate chamber where the steam could cool and condense. The two chambers were connected by a valve that let the steam escape from the boiler. This improved the engine's efficiency—the boiler could stay hot all the time!

A few years later, Watt turned the whole apparatus on its side so that the piston was moving horizontally. He added a slide valve that admitted steam first to one end of the chamber (pushing the piston in one direction) and then to the other end (pushing the piston back). This changed the steam pump into a true steam engine that could drive a locomotive the size of Big Boy!

Explore Other Inventions

▶ Watt's engine helped trigger the Industrial Revolution as many new uses for steam power were found. Find out more about the many other inventors, from tinkerers to engineers, who harnessed the power of steam.

Elements, Compounds, and Mixtures

Pre-Reading Questions

1. What is an element?
2. What is a compound? How are compounds and mixtures different?
3. What are the components of a solution called?

A Groovy Kind of Mixture

When you look at these lamps, you can easily see two different liquids inside them. This mixture is composed of mineral oil, wax, water, and alcohol. The water and alcohol mix, but they remain separated from the globs of wax and oil. In this chapter, you will learn not only about mixtures but also about the elements and compounds that can form mixtures.

Activity

MYSTERY MIXTURE

In this activity, you will separate the different dyes found in an ink mixture.

Procedure

1. Tear a strip of paper (about 3 cm × 15 cm) from a **coffee filter.** Wrap one end of the strip around a **pencil** so that the other end will just touch the bottom of a **clear plastic cup.** Use **tape** to attach the paper to the pencil.

2. Take the paper out of the cup. Using a **water-soluble black marker,** make a small dot in the center of the strip about 2 cm from the bottom.

3. Pour **water** in the cup to a depth of 1 cm. Carefully lower the paper into the cup. Be sure the dot is not under water.

4. Remove the paper when the water is 1 cm from the top. Record your observations in your ScienceLog.

Analysis

5. Infer what happened as the filter paper soaked up the water.

6. Which colors were mixed to make your black ink?

7. Compare your results with those of your classmates. Record your observations.

8. Infer whether the process used to make the ink involved a physical or chemical change. Explain.

Elements

element   nonmetals
pure substance   metalloids
metals

What You'll Do

◆ Describe pure substances.
◆ Describe the characteristics of elements, and give examples.
◆ Explain how elements can be identified.
◆ Classify elements according to their properties.

Imagine you are working as a lab technician for the Break-It-Down Corporation. Your job is to break down materials into the simplest substances you can obtain. One day a material seems particularly difficult to break down. You crush and grind it. You notice that the resulting pieces are smaller, but they are still the same material. You try other physical changes, including melting, boiling, and filtering it, but the material does not change into anything simpler.

Next you try some chemical changes. You pass an electric current through the material but it still does not become any simpler. After recording your observations, you analyze the results of your tests. You then draw a conclusion: the substance must be an element. An **element** is a pure substance that cannot be separated into simpler substances by physical or chemical means, as shown in **Figure 1**.

Figure 1 *No matter what kind of physical or chemical change you attempt, an element cannot be changed into a simpler substance!*

Figure 2 *The atoms of the element iron are alike whether they are in a meteorite or in a common iron skillet.*

An Element Has Only One Type of Particle

A **pure substance** is a substance in which there is only one type of particle. Because elements are pure substances, each element contains only one type of particle. For example, every particle (atom) in a 5 g nugget of the element gold is like every other particle of gold. The particles of a pure substance are alike no matter where that substance is found, as shown in **Figure 2**. Although a meteorite might travel more than 400 million kilometers (about 248 million miles) to reach Earth, the particles of iron in a meteorite are identical to the particles of iron in objects around your home!

Every Element Has a Unique Set of Properties

Each element has a unique set of properties that allows you to identify it. For example, each element has its own *characteristic properties*. These properties do not depend on the amount of material present in a sample of the element. Characteristic properties include some physical properties, such as boiling point, melting point, and density, as well as chemical properties, such as reactivity with acid. The elements helium and krypton are unreactive gases. However, the density (mass per unit volume) of helium is less than the density of air. Therefore, a helium-filled balloon will float up if it is released. Krypton is more dense than air, so a krypton-filled balloon will sink to the ground if it is released.

Identifying Elements by Their Properties Look at the elements cobalt, iron, and nickel, shown in **Figure 3.** Even though these three elements have some similar properties, each can be identified by its unique set of properties.

Notice that the physical properties for the elements in Figure 3 include melting point and density. Other physical properties, such as color, hardness, and texture, could be added to the list. Also, depending on the elements being identified, other chemical properties might be useful. For example, some elements, such as hydrogen and carbon, are flammable. Other elements, such as sodium, react immediately with oxygen. Still other elements, such as zinc, are reactive with acid.

Cobalt

Melting point is 1,495°C.
Density is 8.9 g/cm^3.
Conducts electric current and thermal energy.
Unreactive with oxygen in the air.

Iron

Melting point is 1,535°C.
Density is 7.9 g/cm^3.
Conducts electric current and thermal energy.
Combines slowly with oxygen in the air to form rust.

Nickel

Melting point is 1,455°C.
Density is 8.9 g/cm^3.
Conducts electric current and thermal energy.
Unreactive with oxygen in the air.

Figure 3 *Like all other elements, cobalt, iron, and nickel can be identified by their unique combination of properties.*

Elements Are Classified by Their Properties

Consider how many different breeds of dogs there are. Consider also how you tell one breed from another. Most often you can tell just by their appearance, or what might be called physical properties. **Figure 4** shows several breeds of dogs, which all happen to be terriers. Many terriers are fairly small in size and have short hair. Although not all terriers are exactly alike, they share enough common properties to be classified in the same group.

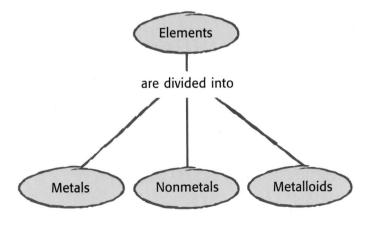

Figure 4 *Even though these dogs are different breeds, they have enough in common to be classified as terriers.*

Elements Are Grouped into Categories

Elements are classified into groups according to their shared properties. Recall the elements iron, nickel, and cobalt. All three are shiny, and all three conduct thermal energy and electric current. Using these shared properties, scientists have grouped these three elements, along with other similar elements, into one large group called metals. Metals are not all exactly alike, but they do have some properties in common.

If You Know the Category, You Know the Properties If you have ever browsed at a music store, you know that the CDs are categorized by type of music. If you like rock-and-roll, you would go to the rock-and-roll section. You might not recognize a particular CD, but you know that it must have the characteristics of rock-and-roll for it to be in this section.

Likewise, you can predict some of the properties of an unfamiliar element by knowing the category to which it belongs. As shown in the concept map in **Figure 5,** elements are classified into three categories—metals, nonmetals, and metalloids. Cobalt, iron, and nickel are classified as metals. If you know that a particular element is a metal, you know that it shares certain properties with iron, nickel, and cobalt. The chart on the next page shows examples of each category and describes the properties that identify elements in each category.

Figure 5 *Elements are divided into three categories: metals, nonmetals, and metalloids.*

Elements

are divided into

Metals   Nonmetals   Metalloids

The Three Major Categories of Elements

Metals

Metals are elements that are shiny and are good conductors of thermal energy and electric current. They are *malleable* (they can be hammered into thin sheets) and *ductile* (they can be drawn into thin wires). Iron has many uses in building and automobile construction. Copper is used in wires and coins.

Lead

Copper

Tin

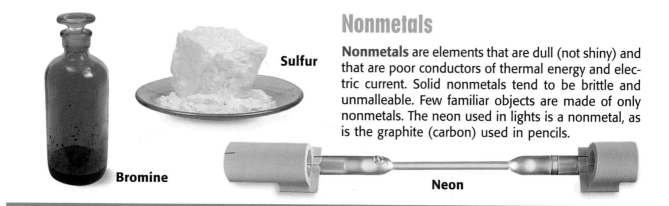

Bromine

Sulfur

Nonmetals

Nonmetals are elements that are dull (not shiny) and that are poor conductors of thermal energy and electric current. Solid nonmetals tend to be brittle and unmalleable. Few familiar objects are made of only nonmetals. The neon used in lights is a nonmetal, as is the graphite (carbon) used in pencils.

Neon

Metalloids

Metalloids, also called semiconductors, are elements that have properties of both metals and nonmetals. Some metalloids are shiny, while others are dull. Metalloids are somewhat malleable and ductile. Some metalloids conduct thermal energy and electric current well. Silicon is used to make computer chips. However, other elements must be added to silicon to make a working chip.

Antimony

Silicon

Boron

SECTION REVIEW

1. What is a pure substance?

2. List three properties that can be used to classify elements.

3. **Applying Concepts** Which category of element would be the least appropriate choice for making a container that can be dropped without shattering? Explain your reasoning.

What You'll Do

♦ Describe the properties of compounds.

♦ Identify the differences between an element and a compound.

♦ Give examples of common compounds.

Compounds

Most elements take part in chemical changes fairly easily, so few elements are found alone in nature. Instead, most elements are found combined with other elements as compounds.

A **compound** is a pure substance composed of two or more elements that are chemically combined. In a compound, a particle is formed when atoms of two or more elements join together. In order for elements to combine, they must *react,* or undergo a chemical change, with one another. In **Figure 6,** you see magnesium reacting with oxygen to form a compound called magnesium oxide. The compound is a new pure substance that is different from the elements that reacted to form it. Most substances you encounter every day are compounds. The table at left lists some familiar examples.

Familiar Compounds

- **table salt—** sodium and chlorine

- **water—** hydrogen and oxygen

- **sugar—** carbon, hydrogen, and oxygen

- **carbon dioxide—** carbon and oxygen

- **baking soda—** sodium, hydrogen, carbon, and oxygen

Figure 6 *As magnesium burns, it reacts with oxygen and forms the compound magnesium oxide.*

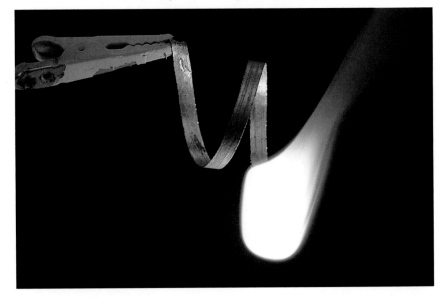

Elements Combine in a Definite Ratio to Form a Compound

Compounds are not random combinations of elements. When a compound forms, the elements join in a specific ratio according to their masses. For example, the ratio of the mass of hydrogen to the mass of oxygen in water is always the same—1 g of hydrogen to 8 g of oxygen. This mass ratio can be written as 1:8 or as the fraction 1/8. Every sample of water has this 1:8 mass ratio of hydrogen to oxygen. If a sample of a compound has a different mass ratio of hydrogen to oxygen, the compound cannot be water.

Every Compound Has a Unique Set of Properties

Each compound has a unique set of properties that allows you to distinguish it from other compounds. Like elements, each compound has its own physical properties, such as boiling point, melting point, density, and color. Compounds can also be identified by their different chemical properties. Some compounds, such as the calcium carbonate found in chalk, react with acid. Others, such as hydrogen peroxide, react when exposed to light. You can see how chemical properties can be used to identify compounds in the QuickLab at right.

A compound has different properties from the elements that form it. Did you know that ordinary table salt is a compound made from two very dangerous elements? Table salt— sodium chloride—consists of sodium (which reacts violently with water) and chlorine (which is poisonous). Together, however, these elements form a harmless compound with unique properties. Take a look at **Figure 7.** Because a compound has different properties from the elements that react to form it, sodium chloride is safe to eat and dissolves (without exploding!) in water.

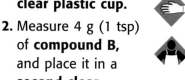

QuickLab

Compound Confusion

1. Measure 4 g (1 tsp) of **compound A,** and place it in a **clear plastic cup.**

2. Measure 4 g (1 tsp) of **compound B,** and place it in a **second clear plastic cup.**

3. Observe the color and texture of each compound. Record your observations.

4. Add 5 mL (1 tsp) of **vinegar** to each cup. Record your observations.

5. Baking soda reacts with vinegar, while powdered sugar does not. Which of these compounds is compound A, and which is compound B?

Figure 7 *Table salt is formed when the elements sodium and chlorine join. The properties of salt are different from the properties of sodium and chlorine.*

Sodium is a soft, silvery white metal that reacts violently with water.

Chlorine is a poisonous, greenish yellow gas.

Sodium chloride, or table salt, is a white solid that dissolves easily in water and is safe to eat.

✔ Self-Check

Do the properties of pure water from a glacier and from a desert oasis differ? *(See page 168 to check your answer.)*

Compounds Can Be Broken Down into Simpler Substances

Some compounds can be broken down into elements through chemical changes. Look at **Figure 8.** When the compound mercury(II) oxide is heated, it breaks down into the elements mercury and oxygen. Likewise, if an electric current is passed through melted table salt, the elements sodium and chlorine are produced.

Other compounds undergo chemical changes to form simpler compounds. These compounds can be broken down into elements through additional chemical changes. For example, carbonic acid is a compound that helps to give carbonated beverages their "fizz," as shown in **Figure 9.** The carbon dioxide and water that are formed can be further broken down into the elements carbon, oxygen, and hydrogen through additional chemical changes.

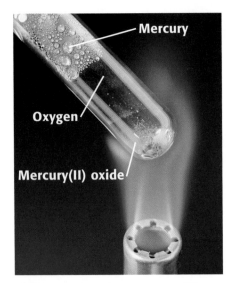

Figure 8 *Heating mercury(II) oxide causes a chemical change that separates it into the elements mercury and oxygen.*

Figure 9 *Opening a carbonated drink can be messy as carbonic acid breaks down into two simpler compounds—carbon dioxide and water.*

Physics

C O N N E C T I O N

The process of using electric current to break compounds into simpler compounds and elements is known as electrolysis. Electrolysis can be used to separate water into hydrogen and oxygen. The elements aluminum and copper and the compound hydrogen peroxide are important industrial products obtained through electrolysis.

Compounds Cannot Be Broken Down by Physical Changes
The only way to break down a compound is through a chemical change. If you pour water through a filter, the water will pass through the filter unchanged. Filtration is a physical change, so it cannot be used to break down a compound. Likewise, a compound cannot be broken down by being ground into a powder or by any other physical process.

Compounds in Your World

You are always surrounded by compounds. Compounds make up the food you eat, the school supplies you use, the clothes you wear—even you!

Compounds in Nature   Proteins are compounds found in all living things. The element nitrogen is needed to make proteins. **Figure 10** shows how some plants get the nitrogen they need. Other plants use nitrogen compounds that are in the soil. Animals get the nitrogen they need by eating plants or by eating animals that have eaten plants. As an animal digests food, the proteins in the food are broken down into smaller compounds that the animal's cells can use.

Another compound that plays an important role in life is carbon dioxide. You exhale carbon dioxide that was made in your body. Plants take in carbon dioxide and use it to make other compounds, including sugar.

Figure 10 *The bumps on the roots of this pea plant are home to bacteria that form compounds from atmospheric nitrogen. The pea plant makes proteins from these compounds.*

Compounds in Industry   The element nitrogen is combined with the element hydrogen to form a compound called ammonia. Ammonia is manufactured for use in fertilizers. Plants can use ammonia as a source of nitrogen for their proteins. Other manufactured compounds are used in medicines, food preservatives, and synthetic fabrics.

The compounds found in nature are usually not the raw materials needed by industry. Often, these compounds must be broken down to provide elements used as raw material. For example, the element aluminum, used in cans, airplanes, and building materials, is not found alone in nature. It is produced by breaking down the compound aluminum oxide.

SECTION REVIEW

1. What is a compound?

2. What type of change is needed to break down a compound?

3. **Analyzing Ideas**   A jar contains samples of the elements carbon and oxygen. Does the jar contain a compound? Explain.

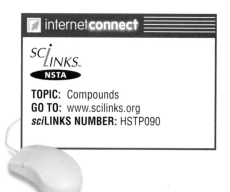

internet**connect**

*sci*LINKS
NSTA

TOPIC: Compounds
GO TO: www.scilinks.org
*sci*LINKS NUMBER: HSTP090

Terms to Learn

mixture   concentration
solution   solubility
solute   suspension
solvent   colloid

What You'll Do

◆ Describe the properties of mixtures.

◆ Describe methods of separating the components of a mixture.

◆ Analyze a solution in terms of its solute, solvent, and concentration.

◆ Compare the properties of solutions, suspensions, and colloids.

Mixtures

Have you ever made your own pizza? You roll out the dough, add a layer of tomato sauce, then add toppings like green peppers, mushrooms, and olives—maybe even some pepperoni! Sprinkle cheese on top, and you're ready to bake. You have just created not only a pizza but also a mixture—and a delicious one at that!

Properties of Mixtures

All mixtures—even pizza—share certain properties. A **mixture** is a combination of two or more substances that are not chemically combined. Two or more materials together form a mixture if they do not react to form a compound. For example, cheese and tomato sauce do not react when they are used to make a pizza.

Figure 11 *Colorless quartz, pink feldspar, and black mica make up the mixture granite.*

Substances in a Mixture Retain Their Identity Because no chemical change occurs, each substance in a mixture has the same chemical makeup it had before the mixture formed. That is, each substance in a mixture keeps its identity. In some mixtures, such as the pizza above or the piece of granite shown in **Figure 11,** you can even see the individual components. In other mixtures, such as salt water, you cannot see all the components.

Mixtures Can Be Physically Separated If you don't like mushrooms on your pizza, you can pick them off. This is a physical change of the mixture. The identities of the substances did not change. In contrast, compounds can be broken down only through chemical changes.

Not all mixtures are as easy to separate as a pizza. You cannot simply pick salt out of a saltwater mixture, but you can separate the salt from the water by heating the mixture. When the water changes from a liquid to a gas, the salt remains behind. Several common techniques for separating mixtures are shown on the following page.

Common Techniques for Separating Mixtures

Distillation is a process that separates a mixture based on the boiling points of the components. Here you see pure water being distilled from a saltwater mixture. In addition to water purification, distillation is used to separate crude oil into its components, such as gasoline and kerosene.

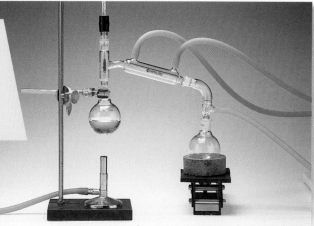

A **magnet** can be used to separate a mixture of the elements iron and aluminum. Iron is attracted to the magnet, but aluminum is not.

The components that make up blood are separated using a machine called a **centrifuge.** This machine separates mixtures according to the densities of the components.

A mixture of the compound sodium chloride (table salt) with the element sulfur requires more than one separation step.

1 The **first step** is to mix them with another compound—water. Salt dissolves in water, but sulfur does not.

2 In the **second step,** the mixture is poured through a filter. The filter traps the solid sulfur.

3 In the **third step,** the sodium chloride is separated from the water by simply evaporating the water.

Mixtures vs. Compounds	
Mixtures	**Compounds**
Components are elements, compounds, or both	Components are elements
Components keep their original properties	Components lose their original properties
Separated by physical means	Separated by chemical means
Formed using any ratio of components	Formed using a set mass ratio of components

The Components of a Mixture Do Not Have a Definite Ratio Recall that a compound has a specific mass ratio of the elements that form it. Unlike compounds, the components of a mixture do not need to be combined in a definite ratio. For example, granite that has a greater amount of feldspar than mica or quartz appears to have a pink color. Granite that has a greater amount of mica than feldspar or quartz appears black. Regardless of which ratio is present, this combination of materials is always a mixture—and it is always called granite.

Air is a mixture composed mostly of nitrogen and oxygen, with smaller amounts of other gases, such as carbon dioxide and water vapor. Some days the air has more water vapor, or is more humid, than on other days. But regardless of the ratio of the components, air is still a mixture. The chart at left summarizes the differences between mixtures and compounds.

SECTION REVIEW

1. What is a mixture?

2. Is a mixture separated by physical or chemical changes?

3. **Applying Concepts** Suggest a procedure to separate iron filings from sawdust. Explain why this procedure works.

BRAIN FOOD

Many substances are soluble in water, including salt, sugar, alcohol, and oxygen. Water does not dissolve everything, but it dissolves so many different solutes that it is often called the universal solvent.

Solutions

A **solution** is a mixture that appears to be a single substance but is composed of particles of two or more substances that are distributed evenly amongst each other. Solutions are often described as *homogeneous mixtures* because they have the same appearance and properties throughout the mixture.

The process in which particles of substances separate and spread evenly throughout a mixture is known as *dissolving*. In solutions, the **solute** is the substance that is dissolved, and the **solvent** is the substance in which the solute is dissolved. A solute is *soluble*, or able to dissolve, in the solvent. A substance that is *insoluble*, or unable to dissolve, forms a mixture that is not homogeneous and therefore is not a solution.

Salt water is a solution. Salt is soluble in water, meaning that salt dissolves in water. Therefore, salt is the solute and water is the solvent. When two liquids or two gases form a solution, the substance with the greater volume is the solvent.

You may think of solutions as being liquids. And, in fact, tap water, soft drinks, gasoline, and many cleaning supplies are liquid solutions. However, solutions may also be gases, such as air, and solids, such as steel. *Alloys* are solid solutions of metals or nonmetals dissolved in metals. Brass is an alloy of the metal zinc dissolved in copper. Steel is an alloy made of the nonmetal carbon and other elements dissolved in iron. Look at the chart below for examples of the different states of matter used as solutes and solvents in solutions.

✔ **Self-Check**

Yellow gold is an alloy made from equal parts copper and silver combined with a greater amount of gold. Identify each component of yellow gold as a solute or solvent. *(See page 168 to check your answer.)*

Examples of Different States in Solutions

Gas in gas	Dry air (oxygen in nitrogen)
Gas in liquid	Soft drinks (carbon dioxide in water)
Liquid in liquid	Antifreeze (alcohol in water)
Solid in liquid	Salt water (salt in water)
Solid in solid	Brass (zinc in copper)

Particles in Solutions Are Extremely Small The particles in solutions are so small that they never settle out, nor can they be filtered out of these mixtures. In fact, the particles are so small, they don't even scatter light. Look at **Figure 12** and see for yourself. The jar on the left contains a solution of sodium chloride in water. The jar on the right contains a mixture of gelatin in water.

Figure 12 *Both of these jars contain mixtures. The mixture in the jar on the left, however, is a solution. The particles in solutions are so small they don't scatter light. Therefore, you can't see the path of light through it.*

÷ 5 ÷ Ω ≤ ∞ +Ω √ 9 ∞ ≤ Σ 2
+

MATH BREAK

Calculating Concentration

Many solutions are colorless. Therefore, you cannot always compare the concentrations of solutions by looking at the color—you have to compare the actual calculated concentrations. One way to calculate the concentration of a liquid solution is to divide the grams of solute by the milliliters of solvent. For example, the concentration of a solution in which 35 g of salt is dissolved in 175 mL of water is

$$\frac{35 \text{ g salt}}{175 \text{ mL water}} = 0.2 \text{ g/mL}$$

Now It's Your Turn

Calculate the concentrations of each solution below. Solution A has 55 g of sugar dissolved in 500 mL of water. Solution B has 36 g of sugar dissolved in 144 mL of water. Which solution is the more dilute one? Which is the more concentrated?

Concentration: How Much Solute Is Dissolved? A measure of the amount of solute dissolved in a solvent is **concentration.** Concentration can be expressed in grams of solute per milliliter of solvent. Knowing the exact concentration of a solution is very important in chemistry and medicine because using the wrong concentration can be dangerous.

Solutions can be described as being *concentrated* or *dilute*. Look at **Figure 13.** Both solutions have the same amount of solvent, but the solution on the left contains less solute than the solution on the right. The solution on the left is dilute while the solution on the right is concentrated. Keep in mind that the terms *concentrated* and *dilute* do not specify the amount of solute that is actually dissolved. Try your hand at calculating concentration and describing solutions as concentrated or dilute in the MathBreak at left.

Figure 13 *The dilute solution on the left contains less solute than the concentrated solution on the right.*

A solution that contains all the solute it can hold at a given temperature is said to be *saturated*. An *unsaturated* solution contains less solute than it can hold at a given temperature. More solute can dissolve in an unsaturated solution.

Solubility: How Much Solute Can Dissolve? If you add too much sugar to a glass of lemonade, not all of the sugar can dissolve. Some of the sugar collects on the bottom of the glass. To determine the maximum amount of sugar that can dissolve, you would need to know the solubility of sugar. The **solubility** of a solute is the amount of solute needed to make a saturated solution using a given amount of solvent at a certain temperature. Solubility is usually expressed in grams of solute per 100 mL of solvent. **Figure 14** on the next page shows the solubility of several different substances in water at different temperatures.

Smelly solutions? Follow your nose and learn more on page 76.

Figure 14 Solubility of Different Substances

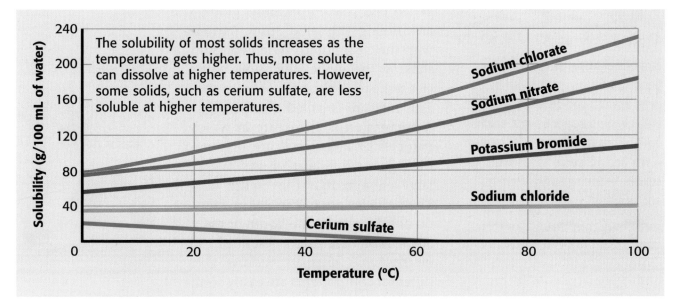

The solubility of most solids increases as the temperature gets higher. Thus, more solute can dissolve at higher temperatures. However, some solids, such as cerium sulfate, are less soluble at higher temperatures.

Y-axis: Solubility (g/100 mL of water) — 40, 80, 120, 160, 200, 240
X-axis: Temperature (°C) — 0, 20, 40, 60, 80, 100

Curves: Sodium chlorate, Sodium nitrate, Potassium bromide, Sodium chloride, Cerium sulfate

Unlike the solubility of most solids in liquids, the solubility of gases in liquids decreases as the temperature is raised. Bubbles of gas appear in hot water long before the water begins to boil. The gases that are dissolved in the water cannot remain dissolved as the temperature increases because the solubility of the gases is lower at higher temperatures.

What Affects How Quickly Solids Dissolve in Liquids?
Many familiar solutions are formed when a solid solute is dissolved in water. Several factors affect how fast the solid will dissolve. Look at **Figure 15** to see three methods used to make a solute dissolve faster. You can see why you will enjoy a glass of lemonade sooner if you stir granulated sugar into the lemonade before adding ice!

Figure 15 *Mixing, heating, and crushing iron(III) chloride increase the speed at which it will dissolve.*

Mixing by stirring or shaking causes the solute particles to separate from one another and spread out more quickly among the solvent particles.

Heating causes particles to move more quickly. The solvent particles can separate the solute particles and spread them out more quickly.

Crushing the solute increases the amount of contact between the solute and the solvent. The particles of solute mix with the solvent more quickly.

Elements, Compounds, and Mixtures **67**

Blood is a suspension. The suspended particles, mainly red blood cells, white blood cells, and platelets, are actually suspended in a solution called plasma. Plasma is 90 percent water and 10 percent dissolved solutes, including sugar, vitamins, and proteins.

Figure 16 *Dirty air is a suspension that could damage a car's engine. The air filter in a car separates dust from air to keep the dust from getting into the engine.*

Suspensions

When you shake up a snow globe, you are mixing the solid snow particles with the clear liquid. When you stop shaking the globe, the snow particles settle to the bottom of the globe. This mixture is called a suspension. A **suspension** is a mixture in which particles of a material are dispersed throughout a liquid or gas but are large enough that they settle out. The particles are insoluble, so they do not dissolve in the liquid or gas. Suspensions are often described as *heterogeneous mixtures* because the different components are easily seen. Other examples of suspensions include muddy water and Italian salad dressing.

The particles in a suspension are fairly large, and they scatter or block light. This often makes a suspension difficult to see through. But the particles are too heavy to remain mixed without being stirred or shaken. If a suspension is allowed to sit undisturbed, the particles will settle out, as in a snow globe.

A suspension can be separated by passing it through a filter. The liquid or gas passes through, but the solid particles are large enough to be trapped by the filter, as shown in **Figure 16.**

APPLY

Shake Well Before Use

Many medicines, such as remedies for upset stomach, are suspensions. The directions on the label instruct you to shake the bottle well before use. Why must you shake the bottle? What problem could arise if you don't?

Colloids

Some mixtures have properties of both solutions and suspensions. These mixtures are known as colloids (KAWL OYDZ). A **colloid** is a mixture in which the particles are dispersed throughout but are not heavy enough to settle out. The particles in a colloid are relatively small and are fairly well mixed. Solids, liquids, and gases can be used to make colloids. You might be surprised at the number of colloids you encounter each day. Milk, mayonnaise, stick deodorant—even the gelatin and whipped cream in **Figure 17**—are colloids. The materials that compose these products do not separate between uses because their particles do not settle out.

Figure 17
This dessert includes two delicious examples of colloids— fruity gelatin and whipped cream.

Although the particles in a colloid are much smaller than the particles in a suspension, they are still large enough to scatter a beam of light shined through the colloid, as shown in **Figure 18.** Finally, unlike a suspension, a colloid cannot be separated by filtration. The particles are small enough to pass through a filter.

Figure 18 *The particles in the colloid fog scatter light, making it difficult for drivers to see the road ahead.*

LabBook

Make a colloid found in your kitchen on page 141 of the LabBook.

SECTION REVIEW

1. List two methods of making a solute dissolve faster.

2. Identify the solute and solvent in a solution made from 15 mL of oxygen and 5 mL of helium.

3. **Comparing Concepts** What are three differences between solutions and suspensions?

📄 internet**connect**

*SCi**LINKS*
NSTA

TOPIC: Mixtures
GO TO: www.scilinks.org
*sci***LINKS NUMBER:** HSTP095

Discovery Lab

Flame Tests

Fireworks make fantastic combinations of color. The colors are the results of burning different compounds. Imagine that you are the lead chemist for a fireworks company. You must identify an unknown compound so that it may be used in the correct fireworks show. You will need to use your knowledge that every compound has a unique set of properties.

MATERIALS

- 4 small test tubes
- test-tube rack
- masking tape
- 4 chloride test solutions
- spark igniter
- Bunsen burner
- wire and holder
- dilute hydrochloric acid in a small beaker
- distilled water in a small beaker

Make a Prediction

1 Can you identify the unknown compound by heating it in a flame? Explain.

Conduct an Experiment

Caution: Be very careful in handling all chemicals. Tell your teacher immediately if you spill a chemical.

2 Arrange the test tubes in the test-tube rack. Use masking tape to label the tubes with the following names: "Calcium chloride," "Potassium chloride," "Sodium chloride," and "Unknown."

3 Copy the table below into your ScienceLog. Then ask your teacher for your portions of the solutions.

Test Results	
Compound	**Color of flame**
Calcium chloride	
Potassium chloride	DO NOT WRITE IN BOOK
Sodium chloride	
Unknown	

4 Light the burner. Clean the wire by first dipping it into the dilute hydrochloric acid. Then dip it into the distilled water. Holding the wooden handle, heat the wire in the blue flame of the burner. Stop heating when the wire is glowing and it no longer colors the flame.
Caution: Use extreme care around an open flame.

Collect Data

5 Dip the clean wire into the first test solution. Hold the wire at the tip of the inner cone of the burner flame. Observe the changes that happen. In the table, record the color given to the flame.

6 Clean the wire by repeating step 4.

7 Repeat steps 5 and 6 for the other solutions.

8 Follow your teacher's instructions for cleanup and disposal.

Analyze the Results

9 Is the flame color a test for the metal or for the chloride in each compound? Explain your answer.

10 What is the identity of your unknown solution? How do you know?

Draw Conclusions

11 Why is it necessary to clean the wire carefully before testing each solution?

12 Infer whether the compound sodium fluoride would make the same color as sodium chloride in a flame test. Explain your answer.

13 Each of the compounds you tested is made from chlorine, which is a poisonous gas at room temperature. Why is it safe to use these compounds without a gas mask?

Chapter Highlights

SECTION 1

Vocabulary

element *(p. 54)*

pure substance *(p. 54)*

metals *(p. 57)*

nonmetals *(p. 57)*

metalloids *(p. 57)*

Section Notes

- A substance in which all the particles are alike is a pure substance.

- An element is a pure substance that cannot be broken down into anything simpler by physical or chemical means.

- Each element has a unique set of physical and chemical properties.

- Elements are classified as metals, nonmetals, or metalloids, based on their properties.

SECTION 2

Vocabulary

compound *(p. 58)*

Section Notes

- A compound is a pure substance composed of two or more elements chemically combined.

- Each compound has a unique set of physical and chemical properties that are different from the properties of the elements that compose it.

- The elements that form a compound always combine in a specific ratio according to their masses.

- Compounds can be broken down into simpler substances by chemical changes.

☑ Skills Check

Math Concepts

CONCENTRATION   The concentration of a solution is a measure of the amount of solute dissolved in a solvent. For example, a solution is formed by dissolving 85 g of sodium nitrate in 170 mL of water. The concentration of the solution is calculated as follows:

$$\frac{85 \text{ g sodium nitrate}}{170 \text{ mL water}} = 0.5 \text{ g/mL}$$

Visual Understanding

THREE CATEGORIES OF ELEMENTS
Elements are classified as metals, nonmetals, or metalloids, based on their properties. The chart on page 57 provides a summary of the properties that distinguish each category.

SEPARATING MIXTURES   Mixtures can be separated through physical changes based on differences in the physical properties of their components. Review the illustrations on page 63 for some techniques for separating mixtures.

Vocabulary

mixture *(p. 62)*

solution *(p. 64)*

solute *(p. 64)*

solvent *(p. 64)*

concentration *(p. 66)*

solubility *(p. 66)*

suspension *(p. 68)*

colloid *(p. 69)*

Section Notes

- A mixture is a combination of two or more substances, each of which keeps its own characteristics.

- Mixtures can be separated by physical means, such as filtration and evaporation.

- The components of a mixture can be mixed in any proportion.

- A solution is a mixture that appears to be a single substance but is composed of a solute dissolved in a solvent. Solutions do not settle, cannot be filtered, and do not scatter light.

- Concentration is a measure of the amount of solute dissolved in a solvent.

- The solubility of a solute is the amount of solute needed to make a saturated solution using a given amount of solvent at a certain temperature.

- Suspensions are heterogeneous mixtures that contain particles large enough to settle out, be filtered, and block or scatter light.

- Colloids are mixtures that contain particles too small to settle out or be filtered but large enough to scatter light.

Labs

A Sugar Cube Race! *(p. 140)*

Making Butter *(p. 141)*

Unpolluting Water *(p. 142)*

---

internet**connect**

GO TO: go.hrw.com

Visit the **HRW** Web site for a variety of learning tools related to this chapter. Just type in the keyword:

KEYWORD: HSTMIX

GO TO: www.scilinks.org

Visit the **National Science Teachers Association** on-line Web site for Internet resources related to this chapter. Just type in the *sci*LINKS number for more information about the topic:

TOPIC: The *Titanic*	*sci*LINKS NUMBER: HSTP080
TOPIC: Elements	*sci*LINKS NUMBER: HSTP085
TOPIC: Compounds	*sci*LINKS NUMBER: HSTP090
TOPIC: Mixtures	*sci*LINKS NUMBER: HSTP095

Chapter Review

Complete the following sentences by choosing the appropriate term from the vocabulary list to fill in each blank:

1. A __?__ has a definite ratio of components.

2. The amount of solute needed to form a saturated solution is the __?__ of the solute.

3. A __?__ can be separated by filtration.

4. A pure substance must be either a(n) __?__ or a(n) __?__.

5. Elements that are brittle and dull are __?__.

6. The substance that dissolves to form a solution is the __?__.

UNDERSTANDING CONCEPTS

Multiple Choice

7. Which of the following increases the solubility of a gas in a liquid?
   a. increasing the temperature
   b. stirring
   c. decreasing the temperature
   d. decreasing the amount of liquid

8. Which of the following best describes chicken noodle soup?
   a. element   c. compound
   b. mixture   d. solution

9. Which of the following does not describe elements?
   a. all the particles are alike
   b. can be broken down into simpler substances
   c. have unique sets of properties
   d. can join together to form compounds

10. A solution that contains a large amount of solute is best described as
    a. unsaturated.   c. dilute.
    b. concentrated.   d. weak.

11. Which of the following substances can be separated into simpler substances only by chemical means?
    a. sodium   c. water
    b. salt water   d. gold

12. Which of the following would not increase the rate at which a solid dissolves?
    a. decreasing the temperature
    b. crushing the solid
    c. stirring
    d. increasing the temperature

13. An element that conducts thermal energy well and is easily shaped is a
    a. metal.
    b. metalloid.
    c. nonmetal.
    d. None of the above

14. In which classification of matter are the components chemically combined?
    a. alloy   c. compound
    b. colloid   d. suspension

Short Answer

15. What is the difference between an element and a compound?

16. When nail polish is dissolved in acetone, which substance is the solute and which is the solvent?

Concept Mapping

17. Use the following terms to create a concept map: matter, element, compound, mixture, solution, suspension, colloid.

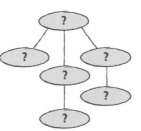

CRITICAL THINKING AND PROBLEM SOLVING

18. Describe a procedure to separate a mixture of salt, finely ground pepper, and pebbles.

19. A light green powder is heated in a test tube. A gas is given off, while the solid becomes black. In which classification of matter does the green powder belong? Explain your reasoning.

20. Why is it desirable to know the exact concentration of solutions rather than whether they are concentrated or dilute?

21. Explain the three properties of mixtures using a fruit salad as an example.

22. To keep the "fizz" in carbonated beverages after they have been opened, should you store them in a refrigerator or in a cabinet? Explain.

MATH IN SCIENCE

23. What is the concentration of a solution prepared by mixing 50 g of salt with 200 mL of water?

24. How many grams of sugar must be dissolved in 150 mL of water to make a solution with a concentration of 0.6 g/mL?

INTERPRETING GRAPHICS

25. Use Figure 14 on page 67 to answer the following questions:
    a. Can 50 g of sodium chloride dissolve in 100 mL of water at 60°C?
    b. How much cerium sulfate is needed to make a saturated solution in 100 mL of water at 30°C?
    c. Is sodium chloride or sodium nitrate more soluble in water at 20°C?

26. Dr. Sol Vent tested the solubility of a compound. The data below was collected using 100 mL of water. Graph Dr. Vent's results. To increase the solubility, would you increase or decrease the temperature? Explain.

Temperature (°C)	10	25	40	60	95
Dissolved solute (g)	150	70	34	25	15

27. What type of mixture is shown in the photo below? Explain.

Reading Check-up

Take a minute to review your answers to the Pre-Reading Questions found at the bottom of page 52. Have your answers changed? If necessary, revise your answers based on what you have learned since you began this chapter.

Science, Technology, and Society

Perfume: Fragrant Solutions

Making perfume is an ancient art. It was practiced, for example, by the ancient Egyptians, who rubbed their bodies with a substance made by soaking fragrant woods and resins in water and oil. From certain references and formulas in the Bible, we know that the ancient Israelites also practiced the art of perfume making. Other sources indicate that this art was also known to the early Chinese, Arabs, Greeks, and Romans.

▲ *Perfumes have been found in the tombs of Egyptians who lived more than 3,000 years ago.*

Only the E-scent-ials

Over time, perfume making has developed into a complicated art. A fine perfume may contain more than 100 different ingredients. The most familiar ingredients come from fragrant plants or flowers, such as sandalwood or roses. These plants get their pleasant odor from their essential oils, which are stored in tiny, baglike parts called sacs. The parts of plants that are used for perfumes include the flowers, roots, and leaves. Other perfume ingredients come from animals and from man-made chemicals.

Making Scents

Perfume makers first remove essential oils from the plants using distillation or reactions with solvents. Then the essential oils are blended with other ingredients to create perfumes. Fixatives, which usually come from animals, make the other odors in the perfume last longer. Oddly enough, most natural fixatives smell awful! For example, civet musk is a foul-smelling liquid that the civet cat sprays on its enemies.

Taking Notes

When you take a whiff from a bottle of perfume, the first odor you detect is called the top note. It is a very fragrant odor that evaporates rather quickly. The middle note, or modifier, adds a different character to the odor of the top note. The base note, or end note, is the odor that lasts the longest.

▲ *Not all perfume ingredients smell good. The foul-smelling oil from the African civet cat is used as a fixative in some perfumes.*

Smell for Yourself

▶ Test a number of different perfumes and colognes to see if you can identify three different notes in each.

Science Fiction

Once upon... in a far away land... there lived a space... who had a ...tic ship ...of silver ...a great haircut that was the gala...

"The Strange Case of Dr. Jekyll and Mr. Hyde"

by Robert Louis Stevenson

A vicious, detestable man murders an old gentleman. A wealthy and respectable scientist commits suicide. Are these two tragedies connected in some way?

Dr. Henry Jekyll is an admirable member of society. He is a doctor and a scientist. Although wild as a young man, Jekyll has become cold and analytical as he has aged and has pursued his scientific theories. Now he wants to understand the nature of human identity. He wants to explore the different parts of the human personality that usually fit together smoothly to make a complete person. His theory is that if he can separate his personality into "good" and "evil" parts, he can get rid of his evil side and lead a happy, useful life. So Jekyll develops a chemical mixture that will allow him to test his theory. The results are startling!

Who is the mysterious Mr. Hyde? He is not a scientist. He is a man of action and anger, who sparks fear in the hearts of those he comes in contact with. Where did he come from? What does he do? How can local residents be protected from his wrath?

Robert Louis Stevenson's story of the decent doctor Henry Jekyll and the violent Edward Hyde is a classic science-fiction story. When Jekyll mixes his "salts" and drinks his chemical mixture, he changes his life—and Edward Hyde's—completely. To find out more, read Stevenson's "The Strange Case of Dr. Jekyll and Mr. Hyde" in the *Holt Anthology of Science Fiction.*

Introduction to Atoms

Pre-Reading Questions

1. What are some ways that scientists have described the atom?

2. What are the parts of the atom, and how are they arranged?

3. How are atoms of all elements alike?

ATOMIC BUBBLES

You probably have made bubbles with a plastic wand and a soapy liquid. To trace the paths of atoms, some scientists also made bubbles, but they did not use a wand. They used a bubble chamber. A bubble chamber is filled with a hot, pressurized liquid that forms bubbles when a charged particle moves through it. Why are scientists interested in bubbles? The bubbles give them information about particles called atoms that make up all objects. In this chapter, you will learn about atoms and experiments that led to the modern atomic theory. You will also learn about the parts and structure of an atom.

START-UP Activity

WHERE IS IT?

Some theories about the internal structure of atoms were formed by observing the effects of aiming very small moving particles at atoms. In this activity, you will form an idea about the location and size of a hidden object by rolling marbles at it.

Procedure

1. Place a **rectangular piece of cardboard** on **four books or blocks** so that each corner of the cardboard rests on a book or block.

2. Ask your teacher to place the **unknown object** under the cardboard. Be sure that you do not see the object.

3. Place a **large piece of paper** on top of the cardboard.

4. Carefully roll a **marble** under the cardboard. Record on the paper the position where the marble enters and exits. Also record the direction it travels.

5. Keep rolling the marble from different directions to collect data about the shape and location of the object.

6. Write down all your observations in your ScienceLog.

Analysis

7. Form a conclusion about the object's shape, size, and location. Record your conclusion in your ScienceLog.

Terms to Learn

atom   model
theory   nucleus
electrons   electron clouds

What You'll Do

◆ Describe some of the experiments that led to the current atomic theory.
◆ Compare the different models of the atom.
◆ Explain how the atomic theory has changed as scientists have discovered new information about the atom.

Development of the Atomic Theory

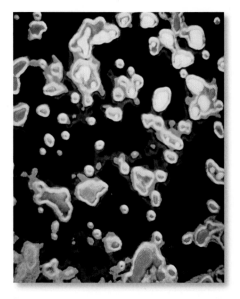

The photo at right shows uranium atoms magnified 3.5 million times by a scanning tunneling microscope. An **atom** is the smallest particle into which an element can be divided and still be the same substance. Atoms make up elements; elements combine to form compounds. Because all matter is made of elements or compounds, atoms are often called the building blocks of matter.

Before the scanning tunneling microscope was invented, in 1981, no one had ever seen an atom. But the existence of atoms is not a new idea. In fact, atomic theory has been around for more than 2,000 years. A **theory** is a unifying explanation for a broad range of hypotheses and observations that have been supported by testing. In this section, you will travel through history to see how our understanding of atoms has developed. Your first stop—ancient Greece.

Democritus Proposes the Atom

Imagine that you cut the silver coin shown in **Figure 1** in half, then cut those halves in half, and so on. Could you keep cutting the pieces in half forever? Around 440 B.C., a Greek philosopher named Democritus (di MAHK ruh tuhs) proposed that you would eventually end up with an "uncuttable" particle. He called this particle an *atom* (from the Greek word *atomos*, meaning "indivisible"). Democritus proposed that all atoms are small, hard particles made of a single material formed into different shapes and sizes. He also claimed that atoms are always moving and that they form different materials by joining together.

Figure 1 *This coin was in use during Democritus's time. Democritus thought the smallest particle in an object like this silver coin was an atom.*

Aristotle Disagrees Aristotle (ER is ᴛᴀʜᴛ uhl), a Greek philosopher who lived from 384 to 322 B.C., disagreed with Democritus's ideas. He believed that you would never end up with an indivisible particle. Although Aristotle's ideas were eventually proved incorrect, he had such a strong influence on popular belief that Democritus's ideas were largely ignored for centuries.

Dalton Creates an Atomic Theory Based on Experiments

By the late 1700s, scientists had learned that elements combine in specific proportions based on mass to form compounds. For example, hydrogen and oxygen always combine in the same proportion to form water. John Dalton, a British chemist and school teacher, wanted to know why. He performed experiments with different substances. His results demonstrated that elements combine in specific proportions because they are made of individual atoms. Dalton, shown in **Figure 2,** published his own atomic theory in 1803. His theory stated the following:

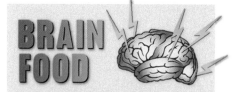

BRAIN FOOD

In 342 or 343 B.C., King Phillip II of Macedon appointed Aristotle to be a tutor for his son, Alexander. Alexander later conquered Greece and the Persian Empire (in what is now Iran) and became known as Alexander the Great.

- **All substances are made of atoms. Atoms are small particles that cannot be created, divided, or destroyed.**

- **Atoms of the same element are exactly alike, and atoms of different elements are different.**

- **Atoms join with other atoms to make new substances.**

Figure 2 *John Dalton developed his atomic theory from observations gathered from many experiments.*

Not Quite Correct Toward the end of the nineteenth century scientists agreed that Dalton's theory explained many of their observations. However, as new information was discovered that could not be explained by Dalton's ideas, the atomic theory was revised to more correctly describe the atom. As you read on, you will learn how Dalton's theory has changed, step by step, into the current atomic theory.

Thomson Finds Electrons in the Atom

In 1897, a British scientist named J. J. Thomson made a discovery that identified an error in Dalton's theory. Using simple equipment (compared with modern equipment), Thomson discovered that there are small particles *inside* the atom. Therefore, atoms *can* be divided into even smaller parts.

Thomson experimented with a cathode-ray tube, as shown in **Figure 3.** He discovered that a positively charged plate (marked with a positive sign in the illustration) attracts the beam. Thomson concluded that the beam was made of particles with a negative electric charge.

Figure 3 Thomson's Cathode-Ray Tube Experiment

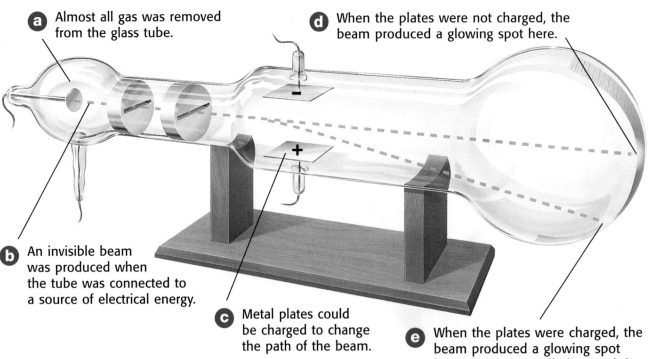

a Almost all gas was removed from the glass tube.

d When the plates were not charged, the beam produced a glowing spot here.

b An invisible beam was produced when the tube was connected to a source of electrical energy.

c Metal plates could be charged to change the path of the beam.

e When the plates were charged, the beam produced a glowing spot here after being pulled toward the positively charged plate.

Just What Is Electric Charge?

Have you ever rubbed a balloon on your hair? The properties of your hair and the balloon seem to change, making them attract one another. To describe these observations, scientists say that the balloon and your hair become "charged." There are two types of electric charges—positive and negative. Objects with opposite charges attract each other, while objects with the same charge push each other away.

Negative Corpuscles Thomson repeated his experiment several times and found that the particle beam behaved in exactly the same way each time. He called the particles in the beam corpuscles (KOR PUHS uhls). His results led him to conclude that corpuscles are present in every type of atom and that all corpuscles are identical. The negatively charged particles found in all atoms are now called **electrons.**

Like Plums in a Pudding Thomson revised the atomic theory to account for the presence of electrons. Because Thomson knew that atoms have no overall charge, he realized that positive charges must be present to balance the negative charges of the electrons. But Thomson didn't know the location of the electrons or of the positive charges. So he proposed a model to describe a possible structure of the atom. A **model** is a representation of an object or system. A model is different from a theory in that a model presents a picture of what the theory explains.

Thomson's model, illustrated in **Figure 4,** came to be known as the plum-pudding model, named for an English dessert that was popular at the time. Today you might call Thomson's model the chocolate-chip-ice-cream model; electrons in the atom could be compared to the chocolate chips found throughout the ice cream!

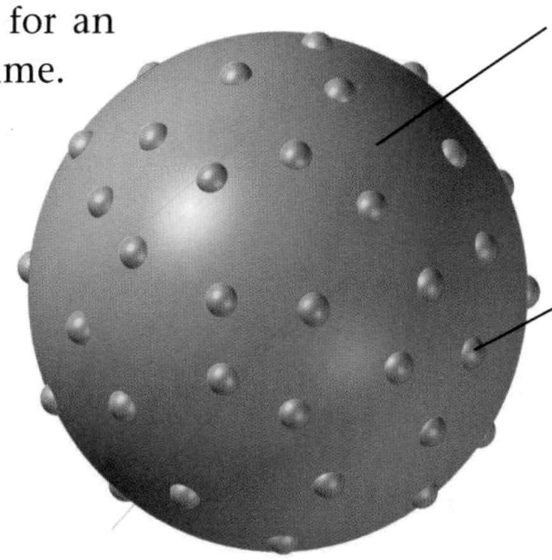

The atom is mostly positively charged material.

Electrons are small, negatively charged particles located throughout the positive material.

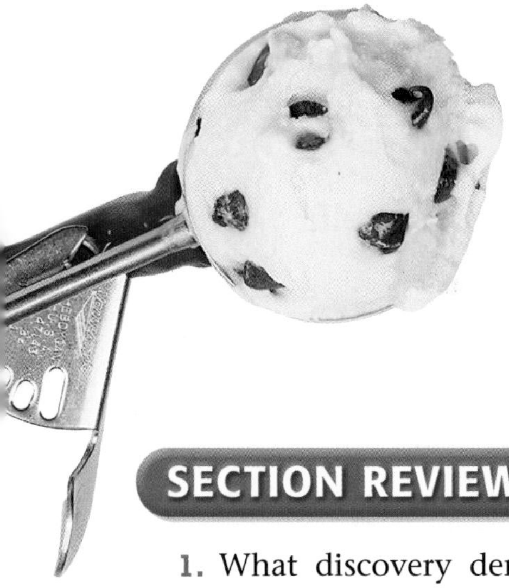

Figure 4 *Thomson's plum-pudding model of the atom is shown above. A modern version of Thomson's model might be chocolate-chip ice cream.*

SECTION REVIEW

1. What discovery demonstrated that atoms are not the smallest particles?

2. What did Dalton do in developing his theory that Democritus did not do?

3. **Analyzing Methods** Why was it important for Thomson to repeat his experiment?

internetconnect

SciLINKS
NSTA

TOPIC: Development of the Atomic Theory
GO TO: www.scilinks.org
*sci*LINKS NUMBER: HSTP255

Rutherford Opens an Atomic "Shooting Gallery"

Find out about Melissa Franklin, a modern atom explorer, on page 101.

In 1909, a former student of Thomson's named Ernest Rutherford decided to test Thomson's theory. He designed an experiment to investigate the structure of the atom. He aimed a beam of small, positively charged particles at a thin sheet of gold foil. These particles were larger than *protons,* even smaller positive particles identified in 1902. **Figure 5** shows a diagram of Rutherford's experiment. To find out where the particles went after being "shot" at the gold foil, Rutherford surrounded the foil with a screen coated with zinc sulfide, a substance that glowed when struck by the particles.

Figure 5 **Rutherford's Gold Foil Experiment**

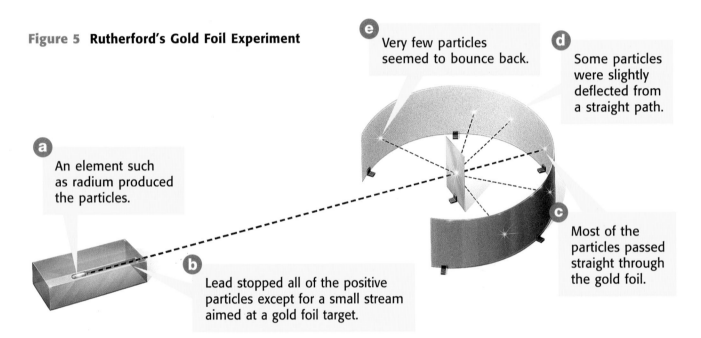

e Very few particles seemed to bounce back.

d Some particles were slightly deflected from a straight path.

a An element such as radium produced the particles.

c Most of the particles passed straight through the gold foil.

b Lead stopped all of the positive particles except for a small stream aimed at a gold foil target.

Rutherford Gets Surprising Results Rutherford thought that if atoms were soft "blobs" of material, as suggested by Thomson, then the particles would pass through the gold and continue in a straight line. Most of the particles did just that. But to Rutherford's great surprise, some of the particles were deflected (turned to one side) a little, some were deflected a great deal, and some particles seemed to bounce back. Rutherford reportedly said,

"It was quite the most incredible event that has ever happened to me in my life. It was almost as if you fired a fifteen-inch shell into a piece of tissue paper and it came back and hit you."

Rutherford Presents a New Atomic Model Rutherford realized that the plum-pudding model of the atom did not explain his results. In 1911, he revised the atomic theory and developed a new model of the atom, as shown in **Figure 6.** To explain the deflection of the particles, Rutherford proposed that in the center of the atom is a tiny, extremely dense, positively charged region called the **nucleus** (NOO klee uhs). Most of the atom's mass is concentrated here. Rutherford reasoned that positively charged particles that passed close by the nucleus were pushed away by the positive charges in the nucleus. A particle that headed straight for a nucleus would be pushed almost straight back in the direction from which it came. From his results, Rutherford calculated that the diameter of the nucleus was 100,000 times smaller than the diameter of the gold atom. To imagine how small this is, look at **Figure 7.**

Figure 6
Rutherford's Model of the Atom

The atom has a small, dense, positively charged **nucleus.**

The atom is mostly **empty space** through which electrons travel.

Electrons travel around the nucleus like planets around the sun, but their exact arrangement could not be described.

Figure 7 *The diameter of this pinhead is 100,000 times smaller than the diameter of the stadium.*

✔ Self-Check

Why did Thomson think the atom contains positive charges? *(See page 168 to check your answer.)*

Bohr States That Electrons Can Jump Between Levels

In 1913, Niels Bohr, a Danish scientist who worked with Rutherford, suggested that electrons travel around the nucleus in definite paths. These paths are located in levels at certain distances from the nucleus, as illustrated in **Figure 8.** Bohr proposed that no paths are located between the levels, but electrons can jump from a path in one level to a path in another level. Think of the levels as rungs on a ladder. You can stand *on* the rungs of a ladder but not *between* the rungs. Bohr's model was a valuable tool in predicting some atomic behavior, but the atomic theory still had room for improvement.

Figure 8 **Bohr's Model of the Atom**

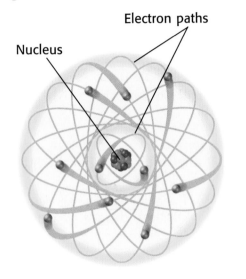

Electron paths

Nucleus

The Modern Theory: Electron Clouds Surround the Nucleus

Many twentieth-century scientists have contributed to our current understanding of the atom. An Austrian physicist named Erwin Schrödinger and a German physicist named Werner Heisenberg made particularly important contributions. Their work further explained the nature of electrons in the atom. For example, electrons do not travel in definite paths as Bohr suggested. In fact, the exact path of a moving electron cannot be predicted. According to the current theory, there are regions inside the atom where electrons are *likely* to be found—these regions are called **electron clouds.** Electron clouds are related to the paths described in Bohr's model. The electron-cloud model of the atom is illustrated in **Figure 9.**

Figure 9 **The Current Model of the Atom**

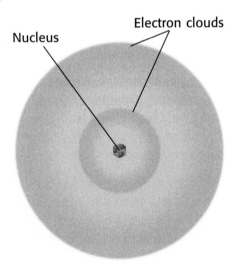

Electron clouds

Nucleus

internetconnect

SciLINKS
NSTA

TOPIC: Modern Atomic Theory
GO TO: www.scilinks.org
*sci*LINKS **NUMBER:** HSTP260

SECTION REVIEW

1. In what part of an atom is most of its mass located?

2. What are two differences between the atomic theory described by Thomson and that described by Rutherford?

3. **Comparing Concepts** Identify the difference in how Bohr's theory and the modern theory describe the location of electrons.

Terms to Learn

protons
atomic mass unit
neutrons
atomic number
isotopes
mass number
atomic mass

What You'll Do

◆ Compare the charge, location, and relative mass of protons, neutrons, and electrons.
◆ Calculate the number of particles in an atom using the atomic number, mass number, and overall charge.
◆ Calculate the atomic mass of elements.

The Atom

In the last section, you learned how the atomic theory developed through centuries of observation and experimentation. Now it's time to learn about the atom itself. In this section, you'll learn about the particles inside the atom, and you'll learn about the forces that act on those particles. But first you'll find out just how small an atom really is.

How Small Is an Atom?

The photograph below shows the pattern that forms when a beam of electrons is directed at a sample of aluminum. By analyzing this pattern, scientists can determine the size of an atom. Analysis of similar patterns for many elements has shown that aluminum atoms, which are average-sized atoms, have a diameter of about 0.00000003 cm. That's three hundred-millionths of a centimeter. That is so small that it would take a stack of 50,000 aluminum atoms to equal the thickness of a sheet of aluminum foil from your kitchen!

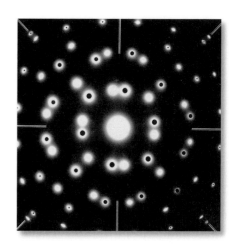

As another example, consider an ordinary penny. Believe it or not, a penny contains about 2×10^{22} atoms, which can be written as 20,000,000,000,000,000,000,000 atoms, of copper and zinc. That's twenty thousand billion billion atoms—over 3,000,000,000,000 times more atoms than there are people on Earth! So if there are that many atoms in a penny, each atom must be very small. You can get a better idea of just how small an atom is in **Figure 10.**

BRAIN FOOD

The size of atoms varies widely. Helium atoms have the smallest diameter, and francium atoms have the largest diameter. In fact, about 600 atoms of helium would fit in the space occupied by a single francium atom!

Figure 10 *If you could enlarge a penny until it was as wide as the continental United States, each of its atoms would be only about 3 cm in diameter—about the size of this table-tennis ball.*

What's Inside an Atom?

As tiny as an atom is, it consists of even smaller particles—protons, neutrons, and electrons—as shown in the model in **Figure 11.** (The particles represented in the figures are not shown in their correct proportions because the electrons would be too small to see.)

Figure 11   Parts of an Atom

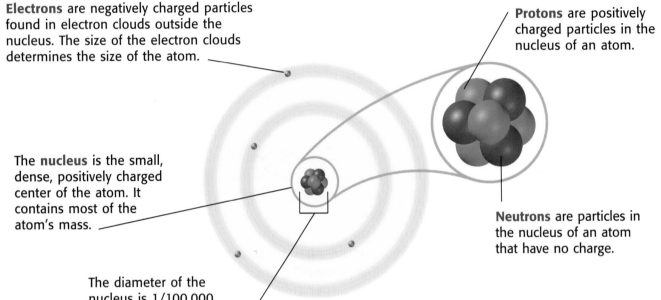

Electrons are negatively charged particles found in electron clouds outside the nucleus. The size of the electron clouds determines the size of the atom.

Protons are positively charged particles in the nucleus of an atom.

The **nucleus** is the small, dense, positively charged center of the atom. It contains most of the atom's mass.

Neutrons are particles in the nucleus of an atom that have no charge.

The diameter of the nucleus is 1/100,000 the diameter of the atom.

The Nucleus Protons are the positively charged particles of the nucleus. It was these particles that repelled Rutherford's "bullets." All protons are identical. The mass of a proton is approximately 1.7×10^{-24} g, which can also be written as 0.00000000000000000000000017 g. Because the masses of particles in atoms are so small, scientists developed a new unit for them. The SI unit used to express the masses of particles in atoms is the **atomic mass unit (amu).** Scientists assign each proton a mass of 1 amu.

Neutrons are the particles of the nucleus that have no charge. All neutrons are identical. Neutrons are slightly more massive than protons, but the difference in mass is so small that neutrons are also given a mass of 1 amu.

Protons and neutrons are the most massive particles in an atom, yet the nucleus has a very small volume. So the nucleus is very dense. If it were possible to have a nucleus the volume of an average grape, that nucleus would have a mass greater than 9 million metric tons!

Proton Profile

Charge: positive
Mass: 1 amu
Location: nucleus

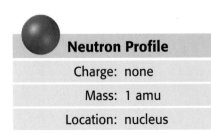

Neutron Profile

Charge: none
Mass: 1 amu
Location: nucleus

Outside of the Nucleus *Electrons* are the negatively charged particles in atoms. Electrons are likely to be found around the nucleus within electron clouds. The charges of protons and electrons are opposite but equal in size. An atom is neutral (has no overall charge) because there are equal numbers of protons and electrons, so their charges cancel out. If the numbers of electrons and protons are not equal, the atom becomes a charged particle called an *ion* (IE ahn). Ions are positively charged if the protons outnumber the electrons, and they are negatively charged if the electrons outnumber the protons.

Electrons are very small in mass compared with protons and neutrons. It takes more than 1,800 electrons to equal the mass of 1 proton. In fact, the mass of an electron is so small that it is usually considered to be zero.

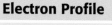

Electron Profile
Charge: negative
Mass: almost zero
Location: electron clouds

SECTION REVIEW

1. What particles form the nucleus?

2. Explain why atoms are neutral.

3. **Summarizing Data** Why do scientists say that most of the mass of an atom is located in the nucleus?

How Do Atoms of Different Elements Differ?

There are over 110 different elements, each of which is made of different atoms. What makes atoms different from each other? To find out, imagine that it's possible to "build" an atom by putting together protons, neutrons, and electrons.

Starting Simply It's easiest to start with the simplest atom. Protons and electrons are found in all atoms, and the simplest atom consists of just one of each. It's so simple it doesn't even have a neutron. Put just one proton in the center of the atom for the nucleus. Then put one electron in the electron cloud, as shown in the model in **Figure 12.** Congratulations! You have just made the simplest atom—a hydrogen atom.

Proton

Electron

Figure 12 *The simplest atom has one proton and one electron.*

Figure 13 *A helium nucleus must have neutrons in it to keep the protons from moving apart.*

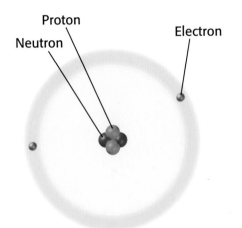

Proton

Neutron

Electron

Now for Some Neutrons Now build an atom containing two protons. Both of the protons are positively charged, so they repel one another. You cannot form a nucleus with them unless you add some neutrons. For this atom, two neutrons will do. Your new atom will also need two electrons outside the nucleus, as shown in the model in **Figure 13.** This is an atom of the element helium.

Building Bigger Atoms You could build a carbon atom using 6 protons, 6 neutrons, and 6 electrons; or you could build an oxygen atom using 8 protons, 9 neutrons, and 8 electrons. You could even build a gold atom with 79 protons, 118 neutrons, and 79 electrons! As you can see, an atom does not have to have equal numbers of protons and neutrons.

The Number of Protons Determines the Element How can you tell which elements these atoms represent? The key is the number of protons. The number of protons in the nucleus of an atom is the **atomic number** of that atom. All atoms of an element have the same atomic number. Every hydrogen atom has only one proton in its nucleus, so hydrogen has an atomic number of 1. Every carbon atom has six protons in its nucleus, so carbon has an atomic number of 6.

Are All Atoms of an Element the Same?

Back in the atom-building workshop, you make an atom that has one proton, one electron, and one neutron, as shown in **Figure 14.** The atomic number of this new atom is 1, so the atom is hydrogen. However, this hydrogen atom's nucleus has two particles; therefore, this atom has a greater mass than the first hydrogen atom you made. What you have is another isotope (IE suh TOHP) of hydrogen. **Isotopes** are atoms that have the same number of protons but have different numbers of neutrons. Atoms that are isotopes of each other are always the same element because the number of protons in each atom is the same.

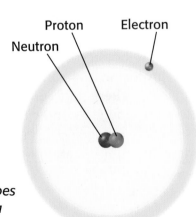

Proton   Electron

Neutron

Figure 14 *The atom in this model and the one in Figure 12 are isotopes because each has one proton but a different number of neutrons.*

Properties of Isotopes Each element has a limited number of isotopes that occur naturally. Some isotopes of an element have unique properties because they are unstable. An unstable atom is an atom whose nucleus can change its composition. This type of isotope is *radioactive*. However, isotopes of an element share most of the same chemical and physical properties. For example, the most common oxygen isotope has 8 neutrons in the nucleus, but other isotopes have 9 or 10 neutrons. All three isotopes are colorless, odorless gases at room temperature. Each isotope has the chemical property of combining with a substance as it burns and even behaves the same in chemical changes in your body.

How Can You Tell One Isotope from Another?

You can identify each isotope of an element by its mass number. The **mass number** is the sum of the protons and neutrons in an atom. Electrons are not included in an atom's mass number because their mass is so small that they have very little effect on the atom's total mass. Look at the boron isotope models shown in **Figure 15** to see how to calculate an atom's mass number.

Isotopes and Light Bulbs

Oxygen reacts, or undergoes a chemical change, with the hot filament in a light bulb, quickly burning out the bulb. Argon does not react with the filament, so a light bulb filled with argon burns out more slowly than one filled with oxygen. Do all three naturally-occurring isotopes of argon have the same effect in light bulbs? Explain your reasoning.

Figure 15 *Each of these boron isotopes has five protons. But because each has a different number of neutrons, each has a different mass number.*

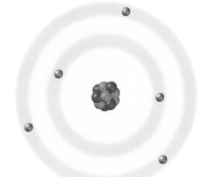

Protons: 5
Neutrons: 5
Electrons: 5
Mass number = protons + neutrons = 10

Protons: 5
Neutrons: 6
Electrons: 5
Mass number = protons + neutrons = 11

Naming Isotopes To identify a specific isotope of an element, write the name of the element followed by a hyphen and the mass number of the isotope. A hydrogen atom with one proton and no neutrons has a mass number of 1. Its name is hydrogen-1. Hydrogen-2 has one proton and one neutron. The carbon isotope with a mass number of 12 is called carbon-12. If you know that the atomic number for carbon is 6, you can calculate the number of neutrons in carbon-12 by subtracting the atomic number from the mass number. For carbon-12, the number of neutrons is 12 − 6, or 6.

12	Mass number
−6	Number of protons (atomic number)
6	Number of neutrons

Calculating the Mass of an Element

Most elements found in nature contain a mixture of two or more stable (nonradioactive) isotopes. For example, all copper is composed of copper-63 atoms and copper-65 atoms. The term *atomic mass* describes the mass of a mixture of isotopes. **Atomic mass** is the weighted average of the masses of all the naturally occurring isotopes of an element. A weighted average accounts for the percentages of each isotope that are present. Copper, including the copper in the Statue of Liberty (shown in **Figure 16**), is 69 percent copper-63 and 31 percent copper-65. The atomic mass of copper is 63.6 amu. You can try your hand at calculating atomic mass by doing the MathBreak at left.

Figure 16 *The copper used to make the Statue of Liberty includes both copper-63 and copper-65. Copper's atomic mass is 63.6 amu.*

What Forces Are at Work in Atoms?

You have seen how atoms are composed of protons, neutrons, and electrons. But what are the *forces* (the pushes or pulls between two objects) acting between these particles? Four basic forces are at work everywhere, including within the atom—gravity, the electromagnetic force, the strong force, and the weak force. These forces are discussed below.

Forces in the Atom

Gravity Probably the most familiar of the four forces is *gravity*. Gravity acts between all objects all the time. The amount of gravity between objects depends on their masses and the distance between them. Gravity pulls objects, such as the sun, Earth, cars, and books, toward one another. However, because the masses of particles in atoms are so small, the force of gravity within atoms is very small.

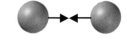

Electromagnetic Force As mentioned earlier, objects that have the same charge repel each other, while objects with opposite charge attract each other. This is due to the *electromagnetic force*. Protons and electrons are attracted to each other because they have opposite charges. The electromagnetic force holds the electrons around the nucleus.

Particles with the same charges repel each other.

Particles with opposite charges attract each other.

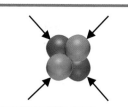

Strong Force Protons push away from one another because of the electromagnetic force. A nucleus containing two or more protons would fly apart if it were not for the *strong force*. At the close distances between protons in the nucleus, the strong force is greater than the electromagnetic force, so the nucleus stays together.

Weak Force The *weak force* is an important force in radioactive atoms. In certain unstable atoms, a neutron can change into a proton and an electron. The weak force plays a key role in this change.

SECTION REVIEW

1. List the charge, location, and mass of a proton, a neutron, and an electron.

2. Determine the number of protons, neutrons, and electrons in an atom of aluminum-27.

3. **Doing Calculations** The metal thallium occurs naturally as 30 percent thallium-203 and 70 percent thallium-205. Calculate the atomic mass of thallium.

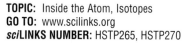

internetconnect

SC*L*INKS.
NSTA

TOPIC: Inside the Atom, Isotopes
GO TO: www.scilinks.org
*sci*LINKS NUMBER: HSTP265, HSTP270

Making Models Lab

Made to Order

Imagine that you are a new worker at the Elements-4-U Company, which makes elements. Your job is to build the atomic nucleus for each element ordered by your customers. You were hired because you know about the makeup of a nucleus and also because you understand how isotopes of an element are different from each other. Now it's time to get to work!

MATERIALS

- 4 protons (white plastic-foam balls, 2–3 cm in diameter)
- 6 neutrons (blue plastic-foam balls, 2–3 cm in diameter)
- 20 connectors (toothpicks)
- periodic table

Procedure

1. Copy the table, shown on the next page, into your ScienceLog. Leave room to add more elements.

2. Before you start the lab, put on your cover goggles. Your first task is to build the nucleus of hydrogen-1. Pick up one proton (a white plastic-foam ball). Congratulations! You have just built a model of a hydrogen-1 nucleus, the simplest nucleus possible.

3. Count the number of protons and neutrons in the nucleus. Fill in rows 1 and 2 for this element in the table.

4. Use the information in rows 1 and 2 to determine the atomic number and mass number of the element. Record this information in the table.

5. Draw a picture of your model in your ScienceLog.

Isotope Data							
	Hydrogen-1	Hydrogen-2	Helium-3	Helium-4	Lithium-7	Beryllium-9	Beryllium-10
No. of protons							
No. of neutrons							
Atomic number							
Mass number							

6 Hydrogen-2 is an isotope of hydrogen that has one proton and one neutron. Using a toothpick, add a neutron to your hydrogen-1 nucleus. (In a nucleus, the protons and neutrons are held together by a force. The toothpicks in this activity stand for the force.) Repeat steps 3–5.

7 Helium-3 is an isotope of helium that has two protons and one neutron. Add one proton to your hydrogen-2 nucleus to create a helium-3 nucleus. Each particle should be connected to the other two particles so that they make a triangle, not a line. Protons and neutrons always form the smallest shape possible. Repeat steps 3–5.

8 For the next part, you will need to use information from the periodic table of the elements. Look at the illustration below. It shows the periodic table entry for carbon. For your job, the most important information in the periodic table is the atomic number. You can find the atomic number of any element at the top of its entry on the table. In the example, the atomic number of carbon is 6.

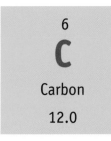

6

C

Carbon

12.0

Atomic number

9 Use the information in the periodic table to build models of the following isotopes of elements: helium-4, lithium-7, beryllium-9, and beryllium-10. Remember to put the protons and neutrons as close together as possible. Each particle should connect to at least two others. Repeat steps 3–5 for each isotope.

Analyze the Results

10 What is the relationship between the number of protons and the atomic number?

11 If you know the atomic number and the mass number of an isotope, how could you figure out the number of neutrons in the nucleus?

12 Look up uranium on the periodic table. What is the atomic number of uranium? How many neutrons does the isotope uranium-235 have?

Communicate Results

13 Compare your model with the models of other groups. How are they the same? How are they different?

Going Further
Working with another group, join your models together. What element (and isotope) have you made?

Chapter Highlights

Vocabulary

atom *(p. 80)*

theory *(p. 80)*

electrons *(p. 83)*

model *(p. 83)*

nucleus *(p. 85)*

electron clouds *(p. 86)*

Section Notes

• Atoms are the smallest particles of an element that retain the properties of the element.

• In ancient Greece, Democritus argued that atoms were the smallest particles in all matter.

• Dalton proposed an atomic theory that stated the following: Atoms are small particles that make up all matter; atoms cannot be created, divided, or destroyed; atoms of an element are exactly alike; atoms of different elements are different; and atoms join together to make new substances.

• Thomson discovered electrons. His plum-pudding model described the atom as a lump of positively charged material with negative electrons scattered throughout.

• Rutherford discovered that atoms contain a small, dense, positively charged center called the nucleus.

• Bohr suggested that electrons move around the nucleus at only certain distances.

• According to the current atomic theory, electron clouds are where electrons are most likely to be in the space around the nucleus.

☑ Skills Check

Math Concepts

ATOMIC MASS The atomic mass of an element takes into account the mass of each isotope and the percentage of the element that exists as that isotope. For example, magnesium occurs naturally as 79 percent magnesium-24, 10 percent magnesium-25, and 11 percent magnesium-26. The atomic mass is calculated as follows:

$$
\begin{array}{rr}
(24 \times 0.79) = & 19.0 \\
(25 \times 0.10) = & 2.5 \\
(26 \times 0.11) = & +\ 2.8 \\
\hline
& 24.3 \ \text{amu}
\end{array}
$$

Visual Understanding

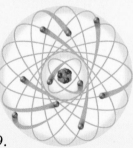

ATOMIC MODELS
The atomic theory has changed over the past several hundred years. To understand the different models of the atom, look over Figures 2, 4, 6, 8, and 9.

PARTS OF THE ATOM Atoms are composed of protons, neutrons, and electrons. To review the particles and their placement in the atom, study Figure 11 on page 88.

Vocabulary

protons *(p. 88)*

atomic mass unit *(p. 88)*

neutrons *(p. 88)*

atomic number *(p. 90)*

isotopes *(p. 90)*

mass number *(p. 91)*

atomic mass *(p. 92)*

Section Notes

- A proton is a positively charged particle with a mass of 1 amu.

- A neutron is a particle with no charge that has a mass of 1 amu.

- An electron is a negatively charged particle with an extremely small mass.

- Protons and neutrons make up the nucleus. Electrons are found in electron clouds outside the nucleus.

- The number of protons in the nucleus of an atom is the atomic number. The atomic number identifies the atoms of a particular element.

- Isotopes of an atom have the same number of protons but have different numbers of neutrons. Isotopes share most of the same chemical and physical properties.

- The mass number of an atom is the sum of the atom's neutrons and protons.

- The atomic mass is an average of the masses of all naturally occurring isotopes of an element.

- The four forces at work in an atom are gravity, the electromagnetic force, the strong force, and the weak force.

 internetconnect

GO TO: go.hrw.com

Visit the **HRW** Web site for a variety of learning tools related to this chapter. Just type in the keyword:

KEYWORD: HSTATS

 **N S T A**

GO TO: www.scilinks.org

Visit the **National Science Teachers Association** on-line Web site for Internet resources related to this chapter. Just type in the *sci*LINKS number for more information about the topic:

TOPIC: Development of the Atomic Theory   *sci*LINKS NUMBER: HSTP255

TOPIC: Modern Atomic Theory   *sci*LINKS NUMBER: HSTP260

TOPIC: Inside the Atom   *sci*LINKS NUMBER: HSTP265

TOPIC: Isotopes   *sci*LINKS NUMBER: HSTP270

Chapter Review

The statements below are false. For each statement, replace the underlined word to make a true statement.

1. Electrons are found in the <u>nucleus</u> of an atom.

2. All atoms of the same element contain the same number of <u>neutrons</u>.

3. <u>Protons</u> have no electric charge.

4. The <u>atomic number</u> of an element is the number of protons and neutrons in the nucleus.

5. The <u>mass number</u> is an average of the masses of all naturally occurring isotopes of an element.

UNDERSTANDING CONCEPTS

Multiple Choice

6. The discovery of which particle proved that the atom is not indivisible?
   a. proton
   b. neutron
   c. electron
   d. nucleus

7. In his gold foil experiment, Rutherford concluded that the atom is mostly empty space with a small, massive, positively charged center because
   a. most of the particles passed straight through the foil.
   b. some particles were slightly deflected.
   c. a few particles bounced back.
   d. All of the above

8. How many protons does an atom with an atomic number of 23 and a mass number of 51 have?
   a. 23
   b. 28
   c. 51
   d. 74

9. An atom has no overall charge if it contains equal numbers of
   a. electrons and protons.
   b. neutrons and protons.
   c. neutrons and electrons.
   d. None of the above

10. Which statement about protons is true?
    a. Protons have a mass of 1/1,840 amu.
    b. Protons have no charge.
    c. Protons are part of the nucleus of an atom.
    d. Protons circle the nucleus of an atom.

11. Which statement about neutrons is true?
    a. Neutrons have a mass of 1 amu.
    b. Neutrons circle the nucleus of an atom.
    c. Neutrons are the only particles that make up the nucleus.
    d. Neutrons have a negative charge.

12. Which of the following determines the identity of an element?
    a. atomic number
    b. mass number
    c. atomic mass
    d. overall charge

13. Isotopes exist because atoms of the same element can have different numbers of
    a. protons.
    b. neutrons.
    c. electrons.
    d. None of the above

Short Answer

14. Why do scientific theories change?

15. What force holds electrons in atoms?

16. In two or three sentences, describe the plum-pudding model of the atom.

Concept Mapping

17. Use the following terms to create a concept map: atom, nucleus, protons, neutrons, electrons, isotopes, atomic number, mass number.

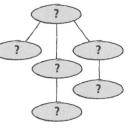

CRITICAL THINKING AND PROBLEM SOLVING

18. Particle accelerators, like the one shown below, are devices that speed up charged particles in order to smash them together. Sometimes the result of the collision is a new nucleus. How can scientists determine whether the nucleus formed is that of a new element or that of a new isotope of a known element?

19. John Dalton made a number of statements about atoms that are now known to be incorrect. Why do you think his atomic theory is still found in science textbooks?

MATH IN SCIENCE

20. Calculate the atomic mass of gallium consisting of 60 percent gallium-69 and 40 percent gallium-71.

21. Calculate the number of protons, neutrons, and electrons in an atom of zirconium-90, which has an atomic number of 40.

INTERPRETING GRAPHICS

22. Study the models below, and answer the questions that follow:

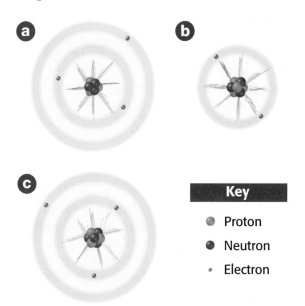

Key
- ● Proton
- ● Neutron
- ° Electron

a. Which models represent isotopes of the same element?

b. What is the atomic number for (a)?

c. What is the mass number for (b)?

23. Predict how the direction of the moving particle in the figure below will change, and explain what causes the change to occur.

Reading Check-up

Take a minute to review your answers to the Pre-Reading Questions found at the bottom of page 78. Have your answers changed? If necessary, revise your answers based on what you have learned since you began this chapter.

Water on the Moon?

When the astronauts of the Apollo space mission explored the surface of the moon in 1969, all they found was rock powder. None of the many samples of moon rocks they carried back to Earth contained any hint of water. Because the astronauts didn't see water on the moon and scientists didn't detect any in the lab, scientists believed there was no water on the moon.

Then in 1994, radio waves suggested another possibility. On a 4-month lunar jaunt, an American spacecraft called *Clementine* beamed radio waves toward various areas of the moon, including a few craters that never receive sunlight. Mostly, the radio waves were reflected by what appeared to be ground-up rock. However, in part of one huge, dark crater, the radio waves were reflected as if by . . . *ice.*

Hunting for Hydrogen Atoms

Scientists were intrigued by *Clementine's* evidence. Two years later, another spacecraft, *Lunar Prospector,* traveled to the moon. Instead of trying to detect water with radio waves, *Prospector* scanned the moon's surface with a device called a *neutron spectrometer* (NS). A neutron spectrometer counts the number of slow neutrons bouncing off a surface. When a neutron hits something about the same mass as itself, it slows down. As it turns out, the only thing close to the mass of a neutron is an *atom* of the lightest of all elements, hydrogen. So when the NS located high concentrations of slow-moving neutrons on the moon, it indicated to scientists that the neutrons were crashing into hydrogen atoms.

As you know, water consists of two atoms of hydrogen and one atom of oxygen. The presence of hydrogen atoms on the moon is more evidence that water may exist there.

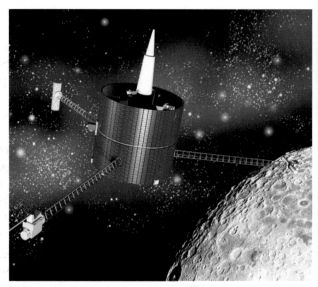

▲ The **Lunar Prospector** *spacecraft may have found water on the moon.*

How Did It Get There?

Some scientists speculate that the water molecules came from comets (which are 90 percent water) that hit the moon more than 4 billion years ago. Water from comets may have landed in the frigid, shadowed craters of the moon, where it mixed with the soil and froze. The Aitken Basin, at the south pole of the moon, where much of the ice was detected, is more than 12 km deep in places. Sunlight never touches most of the crater. And it is very cold—temperatures there may fall to −229°C. The conditions seem right to lock water into place for a very long time.

Think About Lunar Life

▶ Do some research on conditions on the moon. What conditions would humans have to overcome before we could establish a colony there?

EXPERIMENTAL PHYSICIST

In the course of a single day, you could find **Melissa Franklin** operating a huge drill, giving a tour of her lab to a 10-year-old, putting together a gigantic piece of electronic equipment, or even telling a joke. Then you'd see her really get down to business—studying the smallest particles of matter in the universe.

Melissa Franklin is an experimental physicist. "I am trying to understand the forces that describe how everything in the world moves—especially the smallest things," she explains. "I want to find the things that make up all matter in the universe and then try to understand the forces between them."

Other scientists rely on her to test some of the most important hypotheses in physics. For instance, Franklin and her team recently contributed to the discovery of a particle called the top quark. (Quarks are the tiny particles that make up protons and neutrons.)

Physicists had theorized that the top quark might exist but had no evidence. Franklin and more than 450 other scientists worked together to prove the existence of the top quark. Finding it required the use of a massive machine called a particle accelerator. Basically, a particle accelerator smashes particles together, and then scientists look for the remains of the collision. The physicists had to build some very complicated machines to detect the top quark, but the discovery was worth the effort. Franklin and the other researchers have earned the praise of scientists all over the world.

Getting Her Start

"I didn't always want to be a scientist, but what happens is that when you get hooked, you really get hooked. The next thing you know, you're driving forklifts and using overhead cranes while at the same time working on really tiny, incredibly complicated electronics. What I do is a combination of exciting things. It's better than watching TV."

It isn't just the best students who grow up to be scientists. "You can understand the ideas without having to be a math genius," Franklin says. Anyone can have good ideas, she says, absolutely anyone.

Don't Be Shy!

▶ Franklin also has some good advice for young people interested in physics. "Go and bug people at the local university. Just call up a physics person and say, 'Can I come visit you for a couple of hours?' Kids do that with me, and it's really fun." Why don't you give it a try? Prepare for the visit by making a list of questions you would like answered.

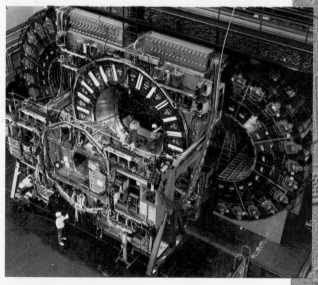

▲ *This particle accelerator was used in the discovery of the top quark.*

The Periodic Table

Pre-Reading Questions

1. How are elements organized in the periodic table?

2. Why is the table of the elements called "periodic"?

3. What one property is shared by elements in a group?

A BUILDING AS A PIECE OF ART!

Would you believe that this strange-looking building is an art museum? It is! It's the Guggenheim Museum in Bilbao, Spain. The building is made of limestone blocks, glass, and the element titanium. Titanium was chosen because it is strong, lightweight, and very resistant to corrosion and rust. In fact, the half-millimeter-thick fish-scale titanium panels covering most of the building are guaranteed to last 100 years! In this chapter, you will learn about some other elements on the periodic table and their properties.

PLACEMENT PATTERN

In this activity, you will determine the pattern behind a new seating chart your teacher has created.

Procedure

1. In your ScienceLog, draw a seating chart for the classroom arrangement given to you by your teacher. Write the name of each of your classmates in the correct place on the chart.

2. Write information about yourself, such as your name, date of birth, hair color, and height, in the space that represents you on the chart.

3. Starting with the people around you, gather the same information about them. Write each person's information in the proper space on the seating chart.

Analysis

4. In your ScienceLog, identify a pattern to the information you gathered that might explain the order of the people in the seating chart. If you cannot find a pattern, collect more information and look again.

5. Test your pattern by gathering information from a person you did not talk to before.

6. If the new information does not support your pattern, reanalyze your data and collect more information to determine another pattern.

Arranging the Elements

Terms to Learn

periodic   period
periodic law   group

What You'll Do

◆ Describe how elements are arranged in the periodic table.
◆ Compare metals, nonmetals, and metalloids based on their properties and on their location in the periodic table.
◆ Describe the difference between a period and a group.

Imagine you go to a new grocery store to buy a box of cereal. You are surprised by what you find. None of the aisles are labeled, and there is no pattern to the products on the shelves! You think it might take you days to find your cereal.

Some scientists probably felt a similar frustration before 1869. By that time, more than 60 elements had been discovered and described. However, it was not until 1869 that the elements were organized in any special way.

Discovering a Pattern

In the 1860s, a Russian chemist named Dmitri Mendeleev began looking for patterns among the properties of the elements. He wrote the names and properties of the elements on pieces of paper. He included density, appearance, atomic mass, melting point, and information about the compounds formed from the element. He then arranged and rearranged the pieces of paper, as shown in **Figure 1.** After much thought and work, he determined that there was a repeating pattern to the properties of the elements when the elements were arranged in order of increasing atomic mass.

Figure 1 *By playing "chemical solitaire" on long train rides, Mendeleev organized the elements according to their properties.*

The Properties of Elements Are Periodic Mendeleev saw that the properties of the elements were **periodic,** meaning they had a regular, repeating pattern. Many things that are familiar to you are periodic. For example, the days of the week are periodic because they repeat in the same order every 7 days.

When the elements were arranged in order of increasing atomic mass, similar chemical and physical properties were observed in every eighth element. Mendeleev's arrangement of the elements came to be known as a periodic table because the properties of the elements change in a periodic way.

Predicting Properties of Missing Elements Look at the section of Mendeleev's periodic table shown in **Figure 2**. Notice the question marks. Mendeleev recognized that there were elements missing and boldly predicted that elements yet to be discovered would fill the gaps. He also predicted the properties of the missing elements by using the pattern of properties in the periodic table. When one of the missing elements, gallium, was discovered a few years later, its properties matched Mendeleev's predictions very well. Since that time, all of the missing elements on Mendeleev's periodic table have been discovered. In the chart below, you can see Mendeleev's predictions for another missing element—germanium—and the actual properties of that element.

```
                              Ni—Co—59
    H—1                           Cu—63,4
          Be—9,4     Mg—24      Zn—65,2
          B—11       Al—27,4     ?—68
          C—12       Si—28       ?—70
          N—14       P—31        As—75
          O—16       S—32        Se—79,4
          F—19       Cl—35,5     Br—80
  Li—7    Na—23      K—39        Rb—85,4
                     Ca—40       Sr—87,6
                      ?—45       Ce—92
                     ?Er—56      La—94
                     ?Yt—60      Di—95
                     ?In—75,6    Th—118?
```

Figure 2 *Mendeleev used question marks to indicate some elements that he believed would later be identified.*

Properties of Germanium		
	Mendeleev's predictions	**Actual properties**
Atomic mass	72	72.6
Density	5.5 g/cm³	5.3 g/cm³
Appearance	dark gray metal	gray metal
Melting point	high melting point	937°C

Changing the Arrangement

Mendeleev noticed that a few elements in the table were not in the correct place according to their properties. He thought that the calculated atomic masses were incorrect and that more accurate atomic masses would eventually be determined. However, new measurements of the atomic masses showed that the masses were in fact correct.

The mystery was solved in 1914 by a British scientist named Henry Moseley (MOHZ lee). From the results of his experiments, Moseley was able to determine the number of protons—the atomic number—in an atom. When he rearranged the elements by atomic number, every element fell into its proper place in an improved periodic table.

Since 1914, more elements have been discovered. Each discovery has supported the periodic law, considered to be the basis of the periodic table. The **periodic law** states that the chemical and physical properties of elements are periodic functions of their atomic numbers. The modern version of the periodic table is shown on the following pages.

Moseley was 26 when he made his discovery. His work allowed him to predict that only three elements were yet to be found between aluminum and gold. The following year, as he fought for the British in World War I, he was killed in action at Gallipoli, Turkey. The British government no longer assigns scientists to combat duty.

Periodic Table of the Elements

Each square on the table includes an element's name, chemical symbol, atomic number, and atomic mass.

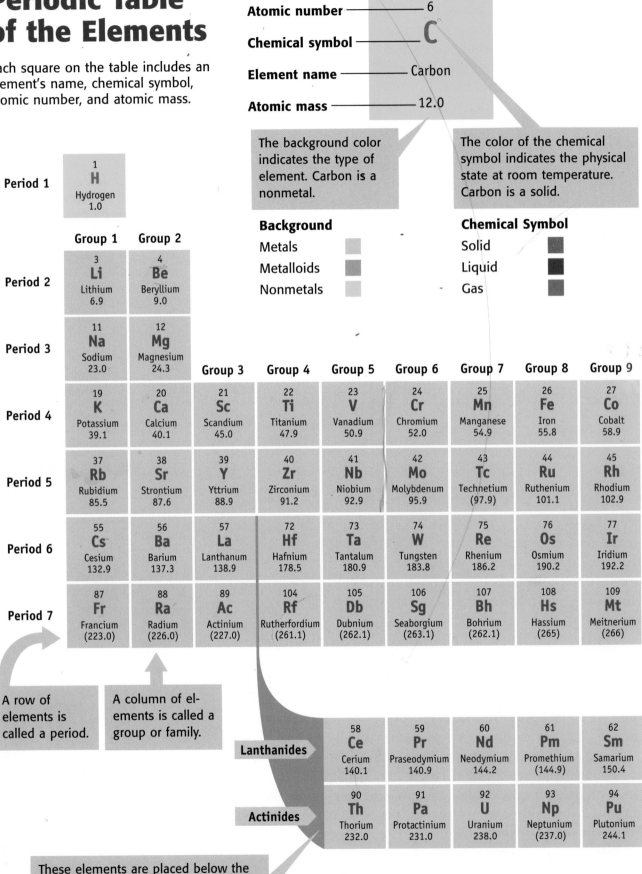

Atomic number ——— 6

Chemical symbol ——— C

Element name ——— Carbon

Atomic mass ——— 12.0

The background color indicates the type of element. Carbon is a nonmetal.

The color of the chemical symbol indicates the physical state at room temperature. Carbon is a solid.

Background
- Metals
- Metalloids
- Nonmetals

Chemical Symbol
- Solid
- Liquid
- Gas

A row of elements is called a period.

A column of elements is called a group or family.

These elements are placed below the table to allow the table to be narrower.

Period 1									
1 H Hydrogen 1.0									

	Group 1	Group 2	Group 3	Group 4	Group 5	Group 6	Group 7	Group 8	Group 9
Period 2	3 Li Lithium 6.9	4 Be Beryllium 9.0							
Period 3	11 Na Sodium 23.0	12 Mg Magnesium 24.3							
Period 4	19 K Potassium 39.1	20 Ca Calcium 40.1	21 Sc Scandium 45.0	22 Ti Titanium 47.9	23 V Vanadium 50.9	24 Cr Chromium 52.0	25 Mn Manganese 54.9	26 Fe Iron 55.8	27 Co Cobalt 58.9
Period 5	37 Rb Rubidium 85.5	38 Sr Strontium 87.6	39 Y Yttrium 88.9	40 Zr Zirconium 91.2	41 Nb Niobium 92.9	42 Mo Molybdenum 95.9	43 Tc Technetium (97.9)	44 Ru Ruthenium 101.1	45 Rh Rhodium 102.9
Period 6	55 Cs Cesium 132.9	56 Ba Barium 137.3	57 La Lanthanum 138.9	72 Hf Hafnium 178.5	73 Ta Tantalum 180.9	74 W Tungsten 183.8	75 Re Rhenium 186.2	76 Os Osmium 190.2	77 Ir Iridium 192.2
Period 7	87 Fr Francium (223.0)	88 Ra Radium (226.0)	89 Ac Actinium (227.0)	104 Rf Rutherfordium (261.1)	105 Db Dubnium (262.1)	106 Sg Seaborgium (263.1)	107 Bh Bohrium (262.1)	108 Hs Hassium (265)	109 Mt Meitnerium (266)

Lanthanides	58 Ce Cerium 140.1	59 Pr Praseodymium 140.9	60 Nd Neodymium 144.2	61 Pm Promethium (144.9)	62 Sm Samarium 150.4
Actinides	90 Th Thorium 232.0	91 Pa Protactinium 231.0	92 U Uranium 238.0	93 Np Neptunium (237.0)	94 Pu Plutonium 244.1

This zigzag line reminds you where the metals, nonmetals, and metalloids are.

Group 10	Group 11	Group 12	Group 13	Group 14	Group 15	Group 16	Group 17	Group 18
								2 **He** Helium 4.0
			5 **B** Boron 10.8	6 **C** Carbon 12.0	7 **N** Nitrogen 14.0	8 **O** Oxygen 16.0	9 **F** Fluorine 19.0	10 **Ne** Neon 20.2
			13 **Al** Aluminum 27.0	14 **Si** Silicon 28.1	15 **P** Phosphorus 31.0	16 **S** Sulfur 32.1	17 **Cl** Chlorine 35.5	18 **Ar** Argon 39.9
28 **Ni** Nickel 58.7	29 **Cu** Copper 63.5	30 **Zn** Zinc 65.4	31 **Ga** Gallium 69.7	32 **Ge** Germanium 72.6	33 **As** Arsenic 74.9	34 **Se** Selenium 79.0	35 **Br** Bromine 79.9	36 **Kr** Krypton 83.8
46 **Pd** Palladium 106.4	47 **Ag** Silver 107.9	48 **Cd** Cadmium 112.4	49 **In** Indium 114.8	50 **Sn** Tin 118.7	51 **Sb** Antimony 121.8	52 **Te** Tellurium 127.6	53 **I** Iodine 126.9	54 **Xe** Xenon 131.3
78 **Pt** Platinum 195.1	79 **Au** Gold 197.0	80 **Hg** Mercury 200.6	81 **Tl** Thallium 204.4	82 **Pb** Lead 207.2	83 **Bi** Bismuth 209.0	84 **Po** Polonium (209.0)	85 **At** Astatine (210.0)	86 **Rn** Radon (222.0)
110 **Uun*** Ununnilium (271)	111 **Uuu*** Unununium (272)	112 **Uub*** Ununbium (277)		114 **Uuq*** Ununquadium (285)		116 **Uuh*** Ununhexium (289)		118 **Uuo*** Ununoctium (293)

A number in parenthesis is the mass number of the most stable form of that element.

63 **Eu** Europium 152.0	64 **Gd** Gadolinium 157.3	65 **Tb** Terbium 158.9	66 **Dy** Dysprosium 162.5	67 **Ho** Holmium 164.9	68 **Er** Erbium 167.3	69 **Tm** Thulium 168.9	70 **Yb** Ytterbium 173.0	71 **Lu** Lutetium 175.0
95 **Am** Americium (243.1)	96 **Cm** Curium (247.1)	97 **Bk** Berkelium (247.1)	98 **Cf** Californium (251.1)	99 **Es** Einsteinium (252.1)	100 **Fm** Fermium (257.1)	101 **Md** Mendelevium (258.1)	102 **No** Nobelium (259.1)	103 **Lr** Lawrencium (262.1)

The official names and symbols for the elements greater than 109 will eventually be approved by a committee of scientists.

Finding Your Way Around the Periodic Table

At first glance, you might think studying the periodic table is like trying to explore a thick jungle without a guide—it would be easy to get lost! However, the table itself contains a lot of information that will help you along the way.

Classes of Elements Elements are classified as metals, nonmetals, and metalloids, according to their properties. The number of electrons in the outer energy level of an atom also helps determine which category an element belongs in. The zigzag line on the periodic table can help you recognize which elements are metals, which are nonmetals, and which are metalloids.

Metals

Most elements are metals. Metals are found to the left of the zigzag line on the periodic table. Atoms of most metals have few electrons in their outer energy level, as shown at right.

Most metals are solid at room temperature. Mercury, however, is a liquid. Some additional information on properties shared by most metals is shown below.

A model of a magnesium atom

Most metals are **good conductors** of thermal energy. This iron griddle conducts thermal energy from a stovetop to cook your favorite foods.

Most metals are **malleable,** meaning that they can be flattened with a hammer without shattering. Aluminum is flattened into sheets to make cans and foil.

Most metals are **ductile,** which means that they can be drawn into thin wires. All metals are good conductors of electric current. The wires in the electrical devices in your home are made from the metal copper.

Metals tend to be **shiny.** You can see a reflection in a mirror because light reflects off the shiny surface of a thin layer of silver behind the glass.

Nonmetals

Nonmetals are found to the right of the zigzag line on the periodic table. Atoms of most nonmetals have an almost complete set of electrons in their outer level, as shown at right. (Atoms of one group of nonmetals, the noble gases, have a complete set of electrons, with most having eight electrons in their outer energy level.)

   More than half of the nonmetals are gases at room temperature. The properties of nonmetals are the opposite of the properties of metals, as shown below.

A model of a chlorine atom

Sulfur, like most nonmetals, is **not shiny.**

Nonmetals are **not malleable or ductile.** In fact, solid nonmetals, like carbon (shown here in the graphite of the pencil lead), are brittle and will break or shatter when hit with a hammer.

Nonmetals are **poor conductors** of thermal energy and electric current. If the gap in a spark plug is too wide, the nonmetals nitrogen and oxygen in the air will stop the spark, and a car's engine will not run.

QuickLab

Conduction Connection

1. Fill a **plastic-foam cup** with **hot water.**
2. Stand a piece of **copper wire** and a **graphite lead** from a mechanical pencil in the water.
3. After 1 minute, touch the top of each object. Record your observations.
4. Which material conducted thermal energy the best? Why?

Metalloids

Metalloids, also called semiconductors, are the elements that border the zigzag line on the periodic table. Atoms of metalloids have about a half-complete set of electrons in their outer energy level, as shown at right.

   Metalloids have some properties of metals and some properties of nonmetals, as shown below.

A model of a silicon atom

Tellurium is **shiny,** but it is also **brittle** and is easily smashed into a powder.

Boron is almost as **hard** as diamond, but it is also **very brittle.** At high temperatures, boron is a good conductor of electric current.

Draw a line down a sheet of paper to divide it into two columns. Look at the elements with atomic numbers 1 through 10 on the periodic table. Write all the chemical symbols and names that follow one pattern in one column on your paper and all chemical symbols and names that follow a second pattern in the second column. Write a sentence describing each pattern you found.

TRY at HOME

Each Element Is Identified by a Chemical Symbol Each square on the periodic table contains information about an element, including its atomic number, atomic mass, name, and chemical symbol. An international committee of scientists is responsible for approving the names and chemical symbols of the elements. The names of the elements come from many sources. For example, some elements are named after important scientists (mendelevium, einsteinium), and others are named for geographical regions (germanium, californium).

The chemical symbol for each element usually consists of one or two letters. The first letter in the symbol is always capitalized, and the second letter, if there is one, is always written in lowercase. The chart below lists the patterns that the chemical symbols follow, and the Activity will help you investigate two of those patterns further.

Writing the Chemical Symbols	
Pattern of chemical symbols	**Examples**
first letter of the name	S—sulfur
first two letters of the name	Ca—calcium
first letter and third or later letter of the name	Mg—magnesium
letter(s) of a word other than the English name	Pb—lead (from the Latin *plumbum,* meaning "lead")
first letter of root words that stand for the atomic number (used for elements whose official names have not yet been chosen)	Uun—ununnilium (uhn uhn NIL ee uhm) (for atomic number 110)

One Set of Symbols

Look at the periodic table shown here. How is it the same as the periodic table you saw earlier? How is it different? Explain why it is important for scientific communication that the chemical symbols used are the same around the world.

元素の周期表

	1 H 1.0079 水素	
1		
2	3 Li 6.941 リチウム	4 Be 9.01218 ベリリウム
3	11 Na 22.98977 ナトリウム	12 Mg 24.305 マグネシウム
4	19 K	20 Ca	21

Rows Are Called Periods Each horizontal row of elements (from left to right) on the periodic table is called a **period.** For example, the row from lithium (Li) to neon (Ne) is Period 2. A row is called a period because the properties of elements in a row follow a repeating, or periodic, pattern as you move across each period. The physical and chemical properties of elements, such as conductivity and the number of electrons in the outer level of atoms, change gradually from those of a metal to those of a nonmetal in each period, as shown in **Figure 3.**

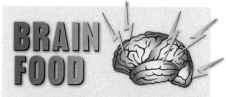

To remember that a period goes from left to right across the periodic table, just think of reading a sentence. You read from left to right across the page until you come to a period.

Figure 3 *The elements in a row become less metallic from left to right.*

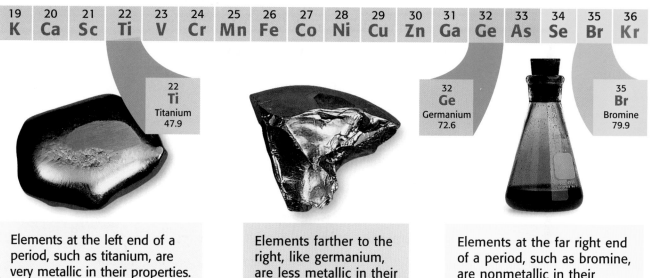

19	20	21	22	23	24	25	26	27	28	29	30	31	32	33	34	35	36
K	Ca	Sc	Ti	V	Cr	Mn	Fe	Co	Ni	Cu	Zn	Ga	Ge	As	Se	Br	Kr

22
Ti
Titanium
47.9

32
Ge
Germanium
72.6

35
Br
Bromine
79.9

Elements at the left end of a period, such as titanium, are very metallic in their properties.

Elements farther to the right, like germanium, are less metallic in their properties.

Elements at the far right end of a period, such as bromine, are nonmetallic in their properties.

Columns Are Called Groups Each column of elements (from top to bottom) on the periodic table is called a **group.** Elements in the same group often have similar chemical and physical properties. For this reason, sometimes a group is also called a family. You will learn more about each group in the next section.

SECTION REVIEW

1. Compare a period and a group on the periodic table.

2. How are the elements arranged in the modern periodic table?

3. **Comparing Concepts** Compare metals, nonmetals, and metalloids in terms of their electrical conductivity.

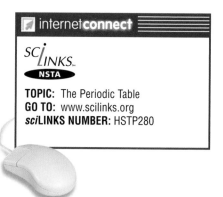

internet**connect**

SC*i*LINKS
NSTA

TOPIC: The Periodic Table
GO TO: www.scilinks.org
*sci*LINKS NUMBER: HSTP280

Terms to Learn

alkali metals
alkaline-earth metals
halogens
noble gases

What You'll Do

◆ Explain why elements in a group often have similar properties.
◆ Describe the properties of the elements in the groups of the periodic table.

Grouping the Elements

You probably know a family with several members that look a lot alike. Or you may have a friend whose little brother or sister acts just like your friend. Members of a family often—but not always—have a similar appearance or behavior. Likewise, the elements in a family or group in the periodic table often—but not always—share similar properties. The properties are similar because the atoms of the elements have the same number of electrons in their outer energy level.

Groups 1 and 2: Very Reactive Metals

The most reactive metals are the elements in Groups 1 and 2. What makes an element reactive? The answer has to do with electrons in the outer energy level of atoms. Atoms will often take, give, or share electrons with other atoms in order to have a complete set of electrons in their outer energy level. Elements whose atoms undergo such processes are *reactive* and combine to form compounds. Elements whose atoms need to take, give, or share only one or two electrons to have a filled outer level tend to be very reactive.

The elements in Groups 1 and 2 are so reactive that they are only found combined with other elements in nature. To study the elements separately, the naturally occurring compounds must first be broken apart through chemical changes.

Group 1: Alkali Metals

Although the element hydrogen appears above the alkali metals on the periodic table, it is not considered a member of Group 1. It will be described separately at the end of this section.

| 3 Li Lithium |
| 11 Na Sodium |
| 19 K Potassium |
| 37 Rb Rubidium |
| 55 Cs Cesium |
| 87 Fr Francium |

Group contains: Metals
Electrons in the outer level: 1
Reactivity: Very reactive
Other shared properties: Soft; silver-colored; shiny; low density

Alkali (AL kuh LIE) **metals** are soft enough to be cut with a knife, as shown in **Figure 4.** The densities of the alkali metals are so low that lithium, sodium, and potassium are actually less dense than water.

Figure 4 *Metals so soft that they can be cut with a knife? Welcome to the alkali metals.*

Alkali metals are the most reactive of the metals. This is because their atoms can easily give away the single electron in their outer level. For example, alkali metals react violently with water, as shown in **Figure 5.** Alkali metals are usually stored in oil to prevent them from reacting with water and oxygen in the atmosphere.

The compounds formed from alkali metals have many uses. Sodium chloride (table salt) can be used to add flavor to your food. Sodium hydroxide can be used to unclog your drains. Potassium bromide is one of several potassium compounds used in photography.

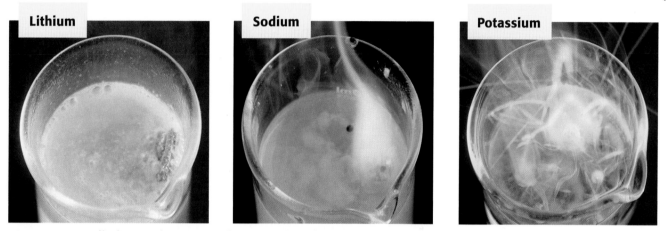

Lithium   **Sodium**   **Potassium**

Figure 5 *As alkali metals react with water, they form hydrogen gas.*

Group 2: Alkaline-earth Metals

| 4
Be
Beryllium |
| 12
Mg
Magnesium |
| 20
Ca
Calcium |
| 38
Sr
Strontium |
| 56
Ba
Barium |
| 88
Ra
Radium |

Group contains: Metals
Electrons in the outer level: 2
Reactivity: Very reactive, but less reactive than alkali metals
Other shared properties: Silver-colored; more dense than alkali metals

Figure 6 *Smile! Calcium, an alkaline-earth metal, is an important component of a compound that makes your bones and teeth healthy.*

Alkaline-earth metals are not as reactive as alkali metals because it is more difficult for atoms to give away two electrons than to give away only one when joining with other atoms.

The alkaline-earth metal magnesium is often mixed with other metals to make low-density materials used in airplanes. Compounds of alkaline-earth metals also have many uses. For example, compounds of calcium are found in cement, plaster, chalk, and even you, as shown in **Figure 6.**

Groups 3–12: Transition Metals

Groups 3–12 do not have individual names. Instead, these groups are described together under the name *transition metals.*

> **Group contains:** Metals
> **Electrons in the outer level:** 1 or 2
> **Reactivity:** Less reactive than alkaline-earth metals
> **Other shared properties:** Shiny; good conductors of thermal energy and electric current; higher densities and melting points (except for mercury) than elements in Groups 1 and 2

21 **Sc** Scandium	22 **Ti** Titanium	23 **V** Vanadium	24 **Cr** Chromium	25 **Mn** Manganese	26 **Fe** Iron	27 **Co** Cobalt	28 **Ni** Nickel	29 **Cu** Copper	30 **Zn** Zinc
39 **Y** Yttrium	40 **Zr** Zirconium	41 **Nb** Niobium	42 **Mo** Molybdenum	43 **Tc** Technetium	44 **Ru** Ruthenium	45 **Rh** Rhodium	46 **Pd** Palladium	47 **Ag** Silver	48 **Cd** Cadmium
57 **La** Lanthanum	72 **Hf** Hafnium	73 **Ta** Tantalum	74 **W** Tungsten	75 **Re** Rhenium	76 **Os** Osmium	77 **Ir** Iridium	78 **Pt** Platinum	79 **Au** Gold	80 **Hg** Mercury
89 **Ac** Actinium	104 **Rf** Rutherfordium	105 **Db** Dubnium	106 **Sg** Seaborgium	107 **Bh** Bohrium	108 **Hs** Hassium	109 **Mt** Meitnerium	110 **Uun** Ununnilium	111 **Uuu** Unununium	112 **Uub** Ununbium

The atoms of transition metals do not give away their electrons as easily as atoms of the Group 1 and Group 2 metals do, making transition metals less reactive than the alkali metals and the alkaline-earth metals. The properties of the transition metals vary widely, as shown in **Figure 7.**

Figure 7 *Transition metals have a wide range of physical and chemical properties.*

Mercury is used in thermometers because, unlike the other transition metals, it is in the liquid state at room temperature.

Some transition metals, including the **titanium** in the artificial hip at right, are not very reactive. But others, such as **iron,** are reactive. The iron in the steel trowel above has reacted with oxygen to form rust.

Many transition metals are silver-colored—but not all! This **gold** ring proves it!

✓ Self-Check

Why are alkali metals more reactive than alkaline-earth metals? *(See page 168 to check your answer.)*

| | 57
La
Lanthanum
138.9 |
| | 89
Ac
Actinium
(227.0) |

Lanthanides and Actinides Some transition metals from Periods 6 and 7 are placed at the bottom of the periodic table to keep the table from being too wide. The properties of the elements in each row tend to be very similar.

	58 **Ce**	59 **Pr**	60 **Nd**	61 **Pm**	62 **Sm**	63 **Eu**	64 **Gd**	65 **Tb**	66 **Dy**	67 **Ho**	68 **Er**	69 **Tm**	70 **Yb**	71 **Lu**
Lanthanides														
Actinides	90 **Th**	91 **Pa**	92 **U**	93 **Np**	94 **Pu**	95 **Am**	96 **Cm**	97 **Bk**	98 **Cf**	99 **Es**	100 **Fm**	101 **Md**	102 **No**	103 **Lr**

Elements in the first row are called *lanthanides* because they follow the transition metal lanthanum. The lanthanides are shiny, reactive metals. Some of these elements are used to make different types of steel. An important use of a compound of one lanthanide element is shown in **Figure 8.**

Elements in the second row are called *actinides* because they follow the transition metal actinium. All atoms of actinides are radioactive, which means they are unstable. The atoms of a radioactive element can change into atoms of a different element. Elements listed after plutonium, element 94, do not occur in nature but are instead produced in laboratories. You might have one of these elements in your home. Very small amounts of americium (AM uhr ISH ee uhm), element 95, are used in some smoke detectors.

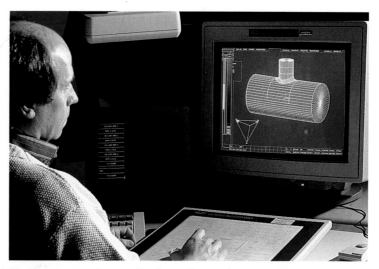

Figure 8 *Seeing red? The color red appears on a computer monitor because of a compound formed from europium that coats the back of the screen.*

SECTION REVIEW

1. What are two properties of the alkali metals?

2. What causes the properties of elements in a group to be similar?

3. **Applying Concepts** Why are neither the alkali metals nor the alkaline-earth metals found uncombined in nature?

internet connect

SC*L*INKS.
NSTA

TOPIC: Metals
GO TO: www.scilinks.org
*sci*LINKS NUMBER: HSTP285

Groups 13–16: Groups with Metalloids

Moving from Group 13 across to Group 16, the elements shift from metals to nonmetals. Along the way, you find the metalloids. These elements have some properties of metals and some properties of nonmetals.

| 5
B
Boron |
| 13
Al
Aluminum |
| 31
Ga
Gallium |
| 49
In
Indium |
| 81
Tl
Thallium |

Group 13: Boron Group

Group contains: One metalloid and four metals
Electrons in the outer level: 3
Reactivity: Reactive
Other shared properties: Solid at room temperature

The most common element from Group 13 is aluminum. In fact, aluminum is the most abundant metal in Earth's crust. Until the 1880s, it was considered a precious metal because the process used to produce pure aluminum was very expensive. In fact, aluminum was even more valuable than gold, as shown in **Figure 9.**

Today, the process is not as difficult or expensive. Aluminum is now an important metal used in making lightweight automobile parts and aircraft, as well as foil, cans, and wires.

Figure 9 *During the 1850s and 1860s, Emperor Napoleon III of France used aluminum dinnerware because aluminum was more valuable than gold!*

Environment
C O N N E C T I O N

Recycling aluminum uses less energy than obtaining aluminum in the first place. Aluminum must be separated from bauxite, a mixture containing naturally occurring compounds of aluminum. Twenty times more electrical energy is required to separate aluminum from bauxite than to recycle used aluminum.

| 6
C
Carbon |
| 14
Si
Silicon |
| 32
Ge
Germanium |
| 50
Sn
Tin |
| 82
Pb
Lead |
| 114
Uuq
Ununquadium |

Group 14: Carbon Group

Group contains: One nonmetal, two metalloids, and two metals
Electrons in the outer level: 4
Reactivity: Varies among the elements
Other shared properties: Solid at room temperature

The metalloids silicon and germanium are used to make computer chips. The metal tin is useful because it is not very reactive. A tin can is really made of steel coated with tin. The tin is less reactive than the steel, and it keeps the steel from rusting.

The nonmetal carbon can be found uncombined in nature, as shown in **Figure 10.** Carbon forms a wide variety of compounds. Some of these compounds, including proteins, fats, and carbohydrates, are essential to life on Earth.

Figure 10 *Diamonds and soot have very different properties, yet both are natural forms of carbon.*

Diamond is the hardest material known. It is used as a jewel and on cutting tools such as saws, drills, and files.

Soot—formed from burning oil, coal, and wood— is used as a pigment in paints and crayons.

Group 15: Nitrogen Group

7	
N	
Nitrogen	
15	
P	
Phosphorus	
33	
As	
Arsenic	
51	
Sb	
Antimony	
83	
Bi	
Bismuth	

Group contains: Two nonmetals, two metalloids, and one metal
Electrons in the outer level: 5
Reactivity: Varies among the elements
Other shared properties: All but nitrogen are solid at room temperature.

Nitrogen, which is a gas at room temperature, makes up about 80 percent of the air you breathe. Nitrogen removed from air is reacted with hydrogen to make ammonia for fertilizers.

Although nitrogen is unreactive, phosphorus is extremely reactive, as shown in **Figure 11.** In fact, phosphorus is only found combined with other elements in nature.

Figure 11
Simply striking a match on the side of this box causes chemicals on the match to react with phosphorus on the box and begin to burn.

Group 16: Oxygen Group

8	
O	
Oxygen	
16	
S	
Sulfur	
34	
Se	
Selenium	
52	
Te	
Tellurium	
84	
Po	
Polonium	
116	
Uuh	
Ununhexium	

Group contains: Three nonmetals, one metalloid, and one metal
Electrons in the outer level: 6
Reactivity: Reactive
Other shared properties: All but oxygen are solid at room temperature.

Oxygen makes up about 20 percent of air. Oxygen is necessary for substances to burn, such as the chemicals on the match in Figure 11. Sulfur, another common member of Group 16, can be found as a yellow solid in nature. The principal use of sulfur is to make sulfuric acid, the most widely used compound in the chemical industry.

FIRES
32 Strike-o
SAFETY MA
Keep Away From
Acme Match Co. Au

Groups 17 and 18: Nonmetals Only

The elements in Groups 17 and 18 are nonmetals. The elements in Group 17 are the most reactive nonmetals, but the elements in Group 18 are the least reactive nonmetals. In fact, the elements in Group 18 normally won't react at all with other elements.

Chlorine is a yellowish green gas.

Bromine is a dark red liquid.

Iodine is a dark gray solid.

Figure 12 *Physical properties of some halogens at room temperature are shown here.*

9
F
Fluorine

17
Cl
Chlorine

35
Br
Bromine

53
I
Iodine

85
At
Astatine

Group 17: Halogens

Group contains: Nonmetals
Electrons in the outer level: 7
Reactivity: Very reactive
Other shared properties: Poor conductors of electric current; react violently with alkali metals to form salts; never found uncombined in nature

Halogens are very reactive nonmetals because their atoms need to gain only one electron to have a complete outer level. The atoms of halogens combine readily with other atoms, especially metals, to gain that missing electron.

Although the chemical properties of the halogens are similar, the physical properties are quite different, as shown in **Figure 12.**

Both chlorine and iodine are used as disinfectants. Chlorine is used to treat water, while iodine mixed with alcohol is used in hospitals.

2
He
Helium

10
Ne
Neon

18
Ar
Argon

36
Kr
Krypton

54
Xe
Xenon

86
Rn
Radon

118
Uuo
Ununoctium

Group 18: Noble Gases

Group contains: Nonmetals
Electrons in the outer level: 8 (2 for helium)
Reactivity: Unreactive
Other shared properties: Colorless, odorless gases at room temperature

Noble gases are unreactive nonmetals. Because the atoms of the elements in this group have a complete set of electrons in their outer level, they do not need to lose or gain any electrons. Therefore, they do not react with other elements under normal conditions.

All of the noble gases are found in Earth's atmosphere in small amounts. Argon, the most abundant noble gas in the atmosphere, makes up almost 1 percent of the atmosphere.

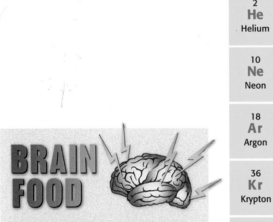

BRAIN FOOD

The term *noble gases* describes the nonreactivity of these elements. Just as nobles, such as kings and queens, did not often mix with common people, the noble gases do not normally react with other elements.

The nonreactivity of the noble gases makes them useful. Ordinary light bulbs last longer when filled with argon than they would if filled with a reactive gas. Because argon is unreactive, it does not react with the metal filament in the light bulb even when the filament gets hot. The low density of helium causes blimps and weather balloons to float, and its nonreactivity makes helium safer to use than hydrogen. One popular use of noble gases that does *not* rely on their nonreactivity is shown in **Figure 13.**

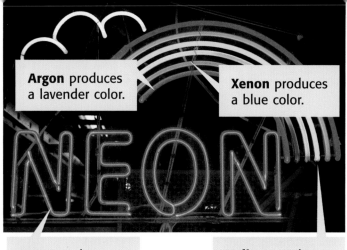

Argon produces a lavender color.

Xenon produces a blue color.

Neon produces an orange-red color.

Helium produces a yellow color.

Figure 13 *Besides neon, other noble gases are often used in "neon" lights.*

Hydrogen Stands Apart

1
H
Hydrogen

Electrons in the outer level: 1
Reactivity: Reactive
Other properties: Colorless, odorless gas at room temperature; low density; reacts explosively with oxygen

The properties of hydrogen do not match the properties of any single group, so hydrogen is set apart from the other elements in the table.

Hydrogen is placed above Group 1 in the periodic table because atoms of the alkali metals also have only one electron in their outer level. Atoms of hydrogen, like atoms of alkali metals, can give away one electron when joining with other atoms. However, hydrogen's physical properties are more like the properties of nonmetals than of metals. As you can see, hydrogen really is in a group of its own.

Hydrogen is the most abundant element in the universe. Hydrogen's reactive nature makes it useful as a fuel in rockets, as shown in **Figure 14.**

Figure 14 *Hydrogen reacts violently with oxygen. The hot water vapor that forms as a result pushes the space shuttle into orbit.*

SECTION REVIEW

1. In which group are the unreactive nonmetals found?

2. What are two properties of the halogens?

3. **Making Predictions** In the future, a new halogen may be synthesized. Predict its atomic number and properties.

4. **Comparing Concepts** Compare the element hydrogen with the alkali metal sodium.

Making Models Lab

Create a Periodic Table

You probably have classification systems for many things in your life, such as your clothes, your books, and your CDs. One of the most important classification systems in science is the periodic table of the elements. In this lab, you will develop your own classification system for a collection of ordinary objects. You will analyze trends in your system and compare your system with the periodic table of the elements.

MATERIALS

- bag of objects
- 20 squares of paper, 3 cm × 3 cm each
- metric balance
- metric ruler
- 2 sheets of graph paper
- computer (optional)

Make Observations

1. Your teacher will give you a bag of objects. It is missing one item. Examine the items carefully. Do you recognize any patterns?

2. Lay out the paper squares on a flat surface so that you have a grid of five rows of four squares each.

3. Analyze the information about the objects to recognize a pattern. Arrange the objects according to the pattern you recognized. You should end up with one blank square for the missing object. In your ScienceLog, describe the basis for your arrangement.

4. Measure the mass (g) and diameter (mm) of each object. Record your results in the appropriate square. Each square except the empty one should have one object placed on it and two measurements written on it.

5. Examine your arrangement again. Does the order in which you arranged your objects still make sense? If necessary, rearrange the squares and their objects to improve your arrangement. Describe the basis for the new arrangement in your ScienceLog.

Form a Hypothesis

6. Based on your observations and your arrangement, form a hypothesis about what you think the identity of the missing object might be. Write your hypothesis in your ScienceLog.

Test the Hypothesis

7 Working across the rows, number the squares 1 to 20. When you get to the end of a row, continue numbering in the first square of the next row.

8 Copy your grid into your ScienceLog, or create a similar grid using a computer. In each square, be sure to list the type of object and label all measurements with appropriate units.

9 On graph paper or on a computer, construct a graph of your data. Show mass on the y-axis and object number on the x-axis. Label each axis, and put a title on the graph.

10 Make a second graph showing diameter on the y-axis and object number on the x-axis. Label each axis, and put a title on the graph.

Analyze the Results

11 Discuss each graph with your classmates. Try to identify any important features of the graph. For example, does the graph form a line or a curve? Is there anything unusual about the graph? What do these features tell you? Write your answers in your ScienceLog.

Draw Conclusions

12 Look back at your hypothesis about the identity of the missing object. Based on your graphs, do you think it is still accurate? Try to improve your description by estimating the mass and diameter of the missing object. Record your estimates in your ScienceLog.

13 How is your arrangement of objects a model of the periodic table of the elements found in this chapter? What are the limitations of your model?

14 How is your experiment similar to the work Mendeleev did with the elements?

Chapter Highlights

SECTION 1

Vocabulary

periodic *(p. 104)*

periodic law *(p. 105)*

period *(p. 111)*

group *(p. 111)*

Section Notes

- Mendeleev developed the first periodic table. He arranged elements in order of increasing atomic mass. The properties of elements repeated in an orderly pattern, allowing Mendeleev to predict properties for elements that had not yet been discovered.

- Moseley rearranged the elements in order of increasing atomic number.

- The periodic law states that the chemical and physical properties of elements are periodic functions of their atomic numbers.

- Elements in the periodic table are divided into metals, metalloids, and nonmetals.

- Each element has a chemical symbol that is recognized around the world.

- A horizontal row of elements is called a period. The elements gradually change from metallic to nonmetallic from left to right across each period.

- A vertical column of elements is called a group or family. Elements in a group usually have similar properties.

☑ Skills Check

Visual Understanding

PERIODIC TABLE OF THE ELEMENTS Scientists rely on the periodic table as a resource for a large amount of information. Review the periodic table on pages 106–107. Pay close attention to the labels and the key; they will help you understand the information presented in the table.

CLASSES OF ELEMENTS Identifying an element as a metal, nonmetal, or metalloid gives you a better idea of the properties of that element. Review the figures on pages 108–109 to understand how to use the zigzag line on the periodic table to identify the classes of elements and to review the properties of elements in each category.

Vocabulary

alkali metals *(p. 112)*

alkaline-earth metals *(p. 113)*

halogens *(p. 118)*

noble gases *(p. 118)*

Section Notes

- The alkali metals (Group 1) are the most reactive metals. Atoms of the alkali metals have one electron in their outer level.

- The alkaline-earth metals (Group 2) are less reactive than the alkali metals. Atoms of the alkaline-earth metals have two electrons in their outer level.

- The transition metals (Groups 3–12) include most of the well-known metals as well as the lanthanides and actinides located below the periodic table.

- Groups 13–16 contain the metalloids along with some metals and nonmetals. The atoms of the elements in each of these groups have the same number of electrons in their outer level.

- The halogens (Group 17) are very reactive nonmetals. Atoms of the halogens have seven electrons in their outer level.

- The noble gases (Group 18) are unreactive nonmetals. Atoms of the noble gases have a complete set of electrons in their outer level.

- Hydrogen is set off by itself because its properties do not match the properties of any one group.

 internet**connect**

GO TO: go.hrw.com

Visit the **HRW** Web site for a variety of learning tools related to this chapter. Just type in the keyword:

KEYWORD: HSTPRT

*SCiLINKS*sm

N S T A

GO TO: www.scilinks.org

Visit the **National Science Teachers Association** on-line Web site for Internet resources related to this chapter. Just type in the *sci*LINKS number for more information about the topic:

TOPIC: The Periodic Table	*sci***LINKS NUMBER:** HSTP280
TOPIC: Metals	*sci***LINKS NUMBER:** HSTP285
TOPIC: Metalloids	*sci***LINKS NUMBER:** HSTP290
TOPIC: Nonmetals	*sci***LINKS NUMBER:** HSTP295
TOPIC: Buckminster Fuller and the Buckyball	*sci***LINKS NUMBER:** HSTP300

Chapter Review

Complete the following sentences by choosing the appropriate term from each pair of terms listed below.

1. Elements in the same vertical column in the periodic table belong to the same ___?___. (*group* or *period*)

2. Elements in the same horizontal row in the periodic table belong to the same ___?___. (*group* or *period*)

3. The most reactive metals are ___?___. (*alkali metals* or *alkaline-earth metals*)

4. Elements that are unreactive are called ___?___. (*noble gases* or *halogens*)

UNDERSTANDING CONCEPTS

Multiple Choice

5. An element that is a very reactive gas is most likely a member of the
   a. noble gases.　　c. halogens.
   b. alkali metals.　　d. actinides.

6. Which statement is true?
   a. Alkali metals are generally found in their uncombined form.
   b. Alkali metals are Group 1 elements.
   c. Alkali metals should be stored under water.
   d. Alkali metals are unreactive.

7. Which statement about the periodic table is false?
   a. There are more metals than nonmetals.
   b. The metalloids are located in Groups 13 through 16.
   c. The elements at the far left of the table are nonmetals.
   d. Elements are arranged by increasing atomic number.

8. One property of most nonmetals is that they are
   a. shiny.
   b. poor conductors of electric current.
   c. flattened when hit with a hammer.
   d. solids at room temperature.

9. Which is a true statement about elements?
   a. Every element occurs naturally.
   b. All elements are found in their uncombined form in nature.
   c. Each element has a unique atomic number.
   d. All of the elements exist in approximately equal quantities.

10. Which is NOT found on the periodic table?
    a. the atomic number of each element
    b. the symbol of each element
    c. the density of each element
    d. the atomic mass of each element

Short Answer

11. Why was Mendeleev's periodic table useful?

12. How is Moseley's basis for arranging the elements different from Mendeleev's?

13. How is the periodic table like a calendar?

14. Describe the location of metals, metalloids, and nonmetals on the periodic table.

Concept Mapping

15. Use the following terms to create a concept map: periodic table, elements, groups, periods, metals, nonmetals, metalloids.

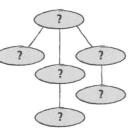

16. When an element with 115 protons in its nucleus is synthesized, will it be a metal, a nonmetal, or a metalloid? Explain.

17. Look at Mendeleev's periodic table in Figure 2. Why was Mendeleev not able to make any predictions about the noble gas elements?

18. Your classmate offers to give you a piece of sodium he found while hiking. What is your response? Explain.

19. Determine the identity of each element described below:
    a. This metal is very reactive, has properties similar to magnesium, and is in the same period as bromine.
    b. This nonmetal is in the same group as lead.
    c. This metal is the most reactive metal in its period and cannot be found uncombined in nature. Each atom of the element contains 19 protons.

MATH IN SCIENCE

20. The chart below shows the percentages of elements in the Earth's crust.

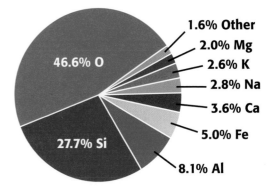

Excluding the "Other" category, what percentage of the Earth's crust is
    a. alkali metals?
    b. alkaline-earth metals?

INTERPRETING GRAPHICS

21. Study the diagram below to determine the pattern of the images. Predict the missing image, and draw it. Identify which properties are periodic and which properties are shared within a group.

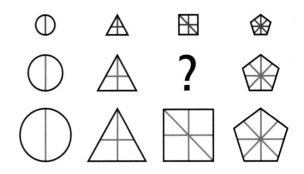

Reading Check-up

Take a minute to review your answers to the Pre-Reading Questions found at the bottom of page 102. Have your answers changed? If necessary, revise your answers based on what you have learned since you began this chapter.

Science, Technology, and Society

The Science of Fireworks

What do the space shuttle and the Fourth of July have in common? The same scientific principles that help scientists launch a space shuttle also help pyrotechnicians create spectacular fireworks shows. The word *pyrotechnics* comes from the Greek words for "fire art." Explosive and dazzling, a fireworks display is both a science and an art.

An Ancient History

More than 1,000 years ago, Chinese civilizations made black powder, the original gunpowder used in pyrotechnics. They used the powder to set off firecrackers and primitive missiles. Black powder is still used today to launch fireworks into the air and to give fireworks an explosive charge. Even the ingredients—saltpeter (potassium nitrate), charcoal, and sulfur—haven't changed since ancient times.

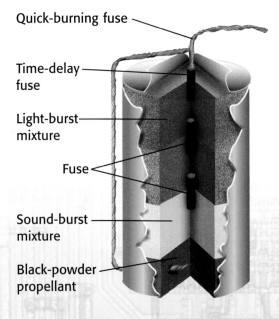

Quick-burning fuse

Time-delay fuse

Light-burst mixture

Fuse

Sound-burst mixture

Black-powder propellant

▲ *Cutaway view of a typical firework. Each shell creates a different type of display.*

Snap, Crackle, Pop!

The shells of fireworks contain the ingredients that create the explosions. Inside the shells, black powder and other chemicals are packed in layers. When ignited, one layer may cause a bright burst of light while a second layer produces a loud booming sound. The shell's shape affects the shape of the explosion. Cylindrical shells produce a trail of lights that looks like an umbrella. Round shells produce a star-burst pattern of lights.

The color and sound of fireworks depend on the chemicals used. To create colors, chemicals like strontium (for red), magnesium (for white), and copper (for blue) can be mixed with the gunpowder.

Explosion in the Sky

Fireworks are launched from metal, plastic, or cardboard tubes. Black powder at the bottom of the shell explodes and shoots the shell into the sky. A fuse begins to burn when the shell is launched. Seconds later, when the explosive chemicals are high in the air, the burning fuse lights another charge of black powder. This ignites the rest of the ingredients in the shell, causing an explosion that lights up the sky!

Bang for Your Buck

▶ The fireworks used during New Year's Eve and Fourth of July celebrations can cost anywhere from $200 to $2,000 apiece. Count the number of explosions at the next fireworks show you see. If each of the fireworks cost just $200 to produce, how much would the fireworks for the entire show cost?

WEIRD SCIENCE

BUCKYBALLS

Researchers are scrambling for the ball—the buckyball, that is. This special form of carbon has 60 carbon atoms linked together in a shape much like a soccer ball. Scientists are having a field day trying to find new uses for this unusual molecule.

Potassium atom trapped inside buckyball

Carbon atoms

Bond

▲ *The buckyball, short for buckminster-fullerene, was named after architect Buckminster Fuller.*

The Starting Lineup

Named for architect Buckminster Fuller, bucky-balls resemble the geodesic domes that are characteristic of the architect's work. Excitement over buckyballs began in 1985 when scientists projected light from a laser onto a piece of graphite. In the soot that remained, researchers found a completely new kind of molecule! Buckyballs are also found in the soot from a candle flame. Some scientists claim to have detected buckyballs in outer space. In fact, one hypothesis suggests that buckyballs might be at the center of the condensing clouds of gas, dust, and debris that form galaxies.

The Game Plan

Ever since buckyballs were discovered, chemists have been busy trying to identify the molecules' properties. One interesting property is that sub-stances can be trapped inside a buckyball. A buckyball can act like a cage that surrounds smaller substances, such as individual atoms. Buckyballs also appear to be both slippery and strong. They can be opened to insert materials, and they can even link together in tubes.

How can buckyballs be used? They may have a variety of uses, from carrying messages through atom-sized wires in computer chips to delivering medicines right where the body needs them. Making tough plastics and cutting tools are uses that are also under investigation. With so many possibilities, scientists expect to get a kick out of bucky-balls for some time!

The Kickoff

▶ A soccer ball is a great model for a buckyball. On the model, the places where three seams meet correspond to the carbon atoms on a buckyball. What represents the bonds between carbon atoms? Does your soccer-ball model have space for all 60 carbon atoms? You'll have to count and see for yourself.

Exploring, inventing, and investigating are essential to the study of science. However, these activities can also be dangerous. To make sure that your experiments and explorations are safe, you must be aware of a variety of safety guidelines.

You have probably heard of the saying, "It is better to be safe than sorry." This is particularly true in a science classroom where experiments and explorations are being performed. Being uninformed and careless can result in serious injuries. Don't take chances with your own safety or with anyone else's.

Following are important guidelines for staying safe in the science classroom. Your teacher may also have safety guidelines and tips that are specific to your classroom and laboratory. Take the time to be safe.

Safety Rules!

Start Out Right

Always get your teacher's permission before attempting any laboratory exploration. Read the procedures carefully, and pay particular attention to safety information and caution statements. If you are unsure about what a safety symbol means, look it up or ask your teacher. You cannot be too careful when it comes to safety. If an accident does occur, inform your teacher immediately, regardless of how minor you think the accident is.

If you are instructed to note the odor of a substance, wave the fumes toward your nose with your hand. Never put your nose close to the source.

Safety Symbols

All of the experiments and investigations in this book and their related worksheets include important safety symbols to alert you to particular safety concerns. Become familiar with these symbols so that when you see them, you will know what they mean and what to do. It is important that you read this entire safety section to learn about specific dangers in the laboratory.

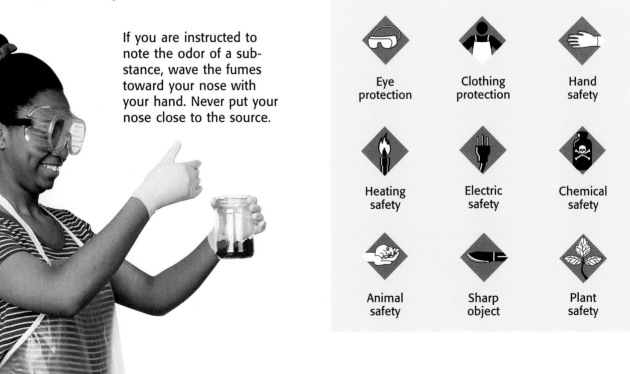

Eye protection	Clothing protection	Hand safety
Heating safety	Electric safety	Chemical safety
Animal safety	Sharp object	Plant safety

Eye Safety

Wear safety goggles when working around chemicals, acids, bases, or any type of flame or heating device. Wear safety goggles any time there is even the slightest chance that harm could come to your eyes. If any substance gets into your eyes, notify your teacher immediately, and flush your eyes with running water for at least 15 minutes. Treat any unknown chemical as if it were a dangerous chemical. Never look directly into the sun. Doing so could cause permanent blindness.

Avoid wearing contact lenses in a laboratory situation. Even if you are wearing safety goggles, chemicals can get between the contact lenses and your eyes. If your doctor requires that you wear contact lenses instead of glasses, wear eye-cup safety goggles in the lab.

Safety Equipment

Know the locations of the nearest fire alarms and any other safety equipment, such as fire blankets and eyewash fountains, as identified by your teacher, and know the procedures for using them.

Be extra careful when using any glassware. When adding a heavy object to a graduated cylinder, tilt the cylinder so the object slides slowly to the bottom.

Neatness

Keep your work area free of all unnecessary books and papers. Tie back long hair, and secure loose sleeves or other loose articles of clothing, such as ties and bows. Remove dangling jewelry. Don't wear open-toed shoes or sandals in the laboratory. Never eat, drink, or apply cosmetics in a laboratory setting. Food, drink, and cosmetics can easily become contaminated with dangerous materials.

Certain hair products (such as aerosol hair spray) are flammable and should not be worn while working near an open flame. Avoid wearing hair spray or hair gel on lab days.

Sharp/Pointed Objects

Use knives and other sharp instruments with extreme care. Never cut objects while holding them in your hands. Place objects on a suitable work surface for cutting.

Heat

Wear safety goggles when using a heating device or a flame. Whenever possible, use an electric hot plate as a heat source instead of an open flame. When heating materials in a test tube, always angle the test tube away from yourself and others. In order to avoid burns, wear heat-resistant gloves whenever instructed to do so.

Chemicals

Wear safety goggles when handling any potentially dangerous chemicals, acids, or bases. If a chemical is unknown, handle it as you would a dangerous chemical. Wear an apron and safety gloves when working with acids or bases or whenever you are told to do so. If a spill gets on your skin or clothing, rinse it off immediately with water for at least 5 minutes while calling to your teacher.

Never mix chemicals unless your teacher tells you to do so. Never taste, touch, or smell chemicals unless you are specifically directed to do so. Before working with a flammable liquid or gas, check for the presence of any source of flame, spark, or heat.

Electricity

Be careful with electrical cords. When using a microscope with a lamp, do not place the cord where it could trip someone. Do not let cords hang over a table edge in a way that could cause equipment to fall if the cord is accidentally pulled. Do not use equipment with damaged cords. Be sure your hands are dry and that the electrical equipment is in the "off" position before plugging it in. Turn off and unplug electrical equipment when you are finished.

Animal Safety

Always obtain your teacher's permission before bringing any animal into the school building. Handle animals only as your teacher directs. Always treat animals carefully and with respect. Wash your hands thoroughly after handling any animal.

Plant Safety

Do not eat any part of a plant or plant seed used in the laboratory. Wash hands thoroughly after handling any part of a plant. When in nature, do not pick any wild plants unless your teacher instructs you to do so.

Glassware

Examine all glassware before use. Be sure that glassware is clean and free of chips and cracks. Report damaged glassware to your teacher. Glass containers used for heating should be made of heat-resistant glass.

Measuring Liquid Volume

In this lab you will use a graduated cylinder to measure and transfer precise amounts of liquids. Remember, in order to accurately measure liquids in a graduated cylinder, you should read the level at the bottom of the meniscus, the curved surface of the liquid.

Procedure

1. Using the masking tape and marker, label the test tubes A, B, C, D, E, and F. Place them in the test-tube rack. Be careful not to confuse the test tubes.

2. Using the 10 mL graduated cylinder and the funnel, pour 14 mL of the red liquid into test tube A. (To do this, first pour 10 mL of the liquid into the test tube and then add 4 mL of liquid.)

3. Rinse the graduated cylinder and funnel between uses.

4. Measure 13 mL of the yellow liquid, and pour it into test tube C. Then measure 13 mL of the blue liquid, and pour it into test tube E.

5. Transfer 4 mL of liquid from test tube C into test tube D. Transfer 7 mL of liquid from test tube E into test tube D.

6. Measure 4 mL of blue liquid from the beaker, and pour it into test tube F. Measure 7 mL of red liquid from the beaker, and pour it into test tube F.

7. Transfer 8 mL of liquid from test tube A into test tube B. Transfer 3 mL of liquid from test tube C into test tube B.

Collect Data

8. Make a data table in your ScienceLog, and record the color of the liquid in each test tube.

9. Use the graduated cylinder to measure the volume of liquid in each test tube, and record the volumes in your data table.

10. Record your color observations in a table of class data prepared by your teacher. Copy the completed table into your ScienceLog.

Analysis

11. Did all of the groups report the same colors? Explain why the colors were the same or different.

12. Why should you not fill the graduated cylinder to the very top?

Materials

- masking tape
- marker
- 6 large test tubes
- test-tube rack
- 10 mL graduated cylinder
- 3 beakers filled with colored liquid
- small funnel

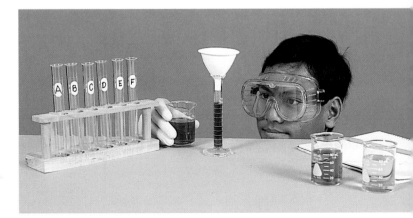

Coin Operated

All pennies are exactly the same, right? Probably not! After all, each penny was made in a certain year at a specific mint, and each has traveled a unique path to reach your classroom. But all pennies *are* similar. In this lab you will investigate differences and similarities among a group of pennies.

Materials

- 10 pennies
- metric balance
- few sheets of paper
- 100 mL graduated cylinder
- water
- paper towels

Procedure

1. Write the numbers 1 through 10 on a page in your ScienceLog, and place a penny next to each number.

2. Use the metric balance to find the mass of each penny to the nearest 0.1 g. Record each measurement in your ScienceLog.

3. On a table that your teacher will provide, make a mark in the correct column of the table for each penny you measured.

4. Separate your pennies into piles based on the class data. Place each pile on its own sheet of paper.

5. Measure and record the mass of each pile. Write the mass on the paper you are using to identify the pile.

6. Fill a graduated cylinder about halfway with water. Carefully measure the volume, and record it.

7. Carefully place the pennies from one pile in the graduated cylinder. Measure and record the new volume.

8. Carefully remove the pennies from the graduated cylinder, and dry them off.

9. Repeat steps 6 through 8 for each pile of pennies.

Analyze the Results

10. Determine the volume of the displaced water by subtracting the initial volume from the final volume. This amount is equal to the volume of the pennies. Record the volume of each pile of pennies.

11. Calculate the density of each pile. To do this, divide the total mass of the pennies by the volume of the pennies. Record the density in your ScienceLog.

Draw Conclusions

12. How is it possible for the pennies to have different densities?

13. What clues might allow you to separate the pennies into the same groups without experimentation? Explain.

Volumania!

You have learned how to measure the volume of a solid object that has square or rectangular sides. But there are lots of objects in the world that have irregular shapes. In this lab activity, you'll learn some ways to find the volume of objects that have irregular shapes.

Part A: Finding the Volume of Small Objects

Procedure

1. Fill a graduated cylinder half full with water. Read the volume of the water, and record it in your ScienceLog. Be sure to look at the surface of the water at eye level and to read the volume at the bottom of the meniscus, as shown below.

Read volume here

2. Carefully slide one of the objects into the tilted graduated cylinder, as shown below.

3. Read the new volume, and record it in your ScienceLog.

4. Subtract the old volume from the new volume. The resulting amount is equal to the volume of the solid object.

5. Use the same method to find the volume of the other objects. Record your results in your ScienceLog.

Analysis

6. What changes do you have to make to the volumes you determine in order to express them correctly?

7. Do the heaviest objects always have the largest volumes? Why or why not?

Materials

Part A
- graduated cylinder
- water
- various small objects supplied by your teacher

Part B
- bottom half of a 2 L plastic bottle or similar container
- water
- aluminum pie pan
- paper towels
- funnel
- graduated cylinder

Part B: Finding the Volume of Your Hand

Procedure

8. Completely fill the container with water. Put the container in the center of the pie pan. Be sure not to spill any of the water into the pie pan.

9. Make a fist, and put your hand into the container up to your wrist.

10. Remove your hand, and let the excess water drip into the container, not the pie pan. Dry your hand with a paper towel.

11. Use the funnel to pour the overflow water into the graduated cylinder. Measure the volume. This is the volume of your hand. Record the volume in your ScienceLog. (Remember to use the correct unit of volume for a solid object.)

12. Repeat this procedure with your other hand.

Analysis

13. Was the volume the same for both of your hands? If not, were you surprised? What might account for a person's hands having different volumes?

14. Would it have made a difference if you had placed your open hand into the container instead of your fist? Explain your reasoning.

15. Compare the volume of your right hand with the volume of your classmates' right hands. Create a class graph of right-hand volumes. What is the average right-hand volume for your class?

Going Further

■ Design an experiment to determine the volume of a person's body. In your plans, be sure to include the materials needed for the experiment and the procedures that must be followed. Include a sketch that shows how your materials and methods would be used in this experiment.

■ Using an encyclopedia, the Internet, or other reference materials, find out how the volumes of very large samples of matter—such as an entire planet—are determined.

Determining Density

The density of an object is its mass divided by its volume. But how does the density of a small amount of a substance relate to the density of a larger amount of the same substance? In this lab, you will calculate the density of one marble and of a group of marbles. Then you will confirm the relationship between the mass and volume of a substance.

Materials

- 100 mL graduated cylinder
- water
- paper towels
- 8 to 10 glass marbles
- metric balance
- graph paper

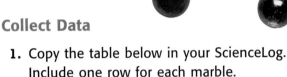

Collect Data

1. Copy the table below in your ScienceLog. Include one row for each marble.

Mass of marble, g	Total mass of marbles, g	Total volume, mL	Volume of marbles, mL (total volume minus 50.0 mL)	Density of marbles, g/mL (total mass of marbles divided by volume of marbles)
DO NOT WRITE IN BOOK			DO NOT WRITE IN BOOK	

2. Fill the graduated cylinder with 50.0 mL of water. If you put in too much water, twist one of the paper towels and use its end to absorb excess water.

3. Measure the mass of a marble as accurately as you can (to at least one-tenth of a gram). Record the marble's mass in the table.

4. Carefully drop the marble in the tilted cylinder, and measure the total volume. Record the volume in the third column.

5. Measure and record the mass of another marble. Add the masses of the marbles together, and record this value in the second column of the table.

6. Carefully drop the second marble in the graduated cylinder. Complete the row of information in the table.

7. Repeat steps 5 and 6, adding one marble at a time. Stop when you run out of marbles, the water no longer completely covers the marbles, or the graduated cylinder is full.

Analyze the Results

8. Examine the data in your table. As the number of marbles increases, what happens to the total mass of the marbles? What happens to the volume of the marbles? What happens to the density of the marbles?

9. Graph the total mass of the marbles (y-axis) versus the volume of the marbles (x-axis). Is the graph a straight line or a curved line?

Draw Conclusions

10. Does the density of a substance depend on the amount of substance present? Explain how your results support your answer.

Going Further

Calculate the slope of the graph. How does the slope compare with the values in the column titled "Density of marbles"? Explain.

DISCOVERY LAB

Layering Liquids

You have learned that liquids form layers according to their densities. In this lab, you'll discover whether it matters in which order you add the liquids.

Make a Prediction

1. Does the order in which you add liquids of different densities to a container affect the order of the layers formed by those liquids?

Conduct an Experiment

2. Using the graduated cylinders, add 10 mL of each liquid to the clear container. Remember to read the volume at the bottom of the meniscus, as shown below. In your ScienceLog, record the order in which you added the liquids.

3. Observe the liquids in the container. In your ScienceLog, sketch what you see. Be sure to label the layers and the colors.

4. Add 10 mL more of liquid C. Observe what happens, and write your observations in your ScienceLog.

5. Add 20 mL more of liquid A. Observe what happens, and write your observations in your ScienceLog.

Analyze Your Results

6. Which of the liquids has the greatest density? Which has the least density? How can you tell?

7. Did the layers change position when you added more of liquid C? Explain your answer.

8. Did the layers change position when you added more of liquid A? Explain your answer.

Materials

- liquid A
- liquid B
- liquid C
- beaker or other small, clear container
- 10 mL graduated cylinders (3)
- 3 funnels

Communicate Your Results

9. Find out in what order your classmates added the liquids to the container. Compare your results with those of a classmate who added the liquids in a different order. Were your results different? In your ScienceLog, explain why or why not.

Draw Conclusions

10. Based on your results, evaluate your prediction from step 1.

Full of Hot Air!

Why do hot-air balloons float gracefully above Earth, while balloons you blow up fall to the ground? The answer has to do with the density of the air inside the balloon. Density is mass per unit volume, and volume is affected by changes in temperature. In this experiment, you will investigate the relationship between the temperature of a gas and its volume. Then you will be able to determine how the temperature of a gas affects its density.

Materials

- 2 aluminum pans
- water
- metric ruler
- hot plate
- ice water
- balloon
- 250 mL beaker
- heat-resistant gloves

Form a Hypothesis

1. How does an increase or decrease in temperature affect the volume of a balloon? Write your hypothesis in your ScienceLog.

Test the Hypothesis

2. Fill an aluminum pan with water about 4 to 5 cm deep. Put the pan on the hot plate, and turn the hot plate on.

3. While the water is heating, fill the other pan 4 to 5 cm deep with ice water.

4. Blow up a balloon inside the 500 mL beaker, as shown. The balloon should fill the beaker but should not extend outside the beaker. Tie the balloon at its opening.

5. Place the beaker and balloon in the ice water. Observe what happens. Record your observations in your ScienceLog.

6. Remove the balloon and beaker from the ice water. Observe the balloon for several minutes. Record any changes.

7. Put on heat-resistant gloves. When the hot water begins to boil, put the beaker and balloon in the hot water. Observe the balloon for several minutes, and record your observations.

8. Turn off the hot plate. When the water has cooled, carefully pour it into a sink.

Analyze the Results

9. Summarize your observations of the balloon. Relate your observations to Charles's law.

10. Was your hypothesis for step 1 supported? If not, revise your hypothesis.

Draw Conclusions

11. Based on your observations, how is the density of a gas affected by an increase or decrease in temperature?

12. Explain in terms of density and Charles's law why heating the air allows a hot-air balloon to float.

Can Crusher

Condensation can occur when gas particles come near the surface of a liquid. The gas particles slow down because they are attracted to the liquid. This reduction in speed causes the gas particles to condense into a liquid. In this lab, you'll see that particles that have condensed into a liquid don't take up as much space and therefore don't exert as much pressure as they did in the gaseous state.

Materials

- water
- 2 empty aluminum cans
- heat-resistant gloves
- hot plate
- tongs
- 1 L beaker

Conduct an Experiment

1. Place just enough water in an aluminum can to slightly cover the bottom.

2. Put on heat-resistant gloves. Place the aluminum can on a hot plate turned to the highest temperature setting.

3. Heat the can until the water is boiling. Steam should be rising vigorously from the top of the can.

4. Using tongs, quickly pick up the can and place the top 2 cm of the can upside down in the 1 L beaker filled with room-temperature water.

5. Describe your observations in your ScienceLog.

Analyze the Results

6. The can was crushed because the atmospheric pressure outside the can became greater than the pressure inside the can. Explain what happened inside the can to cause this.

Draw Conclusions

7. Inside every popcorn kernel is a small amount of water. When you make popcorn, the water inside the kernels is heated until it becomes steam. Explain how the popping of the kernels is the opposite of what you saw in this lab. Be sure to address the effects of pressure in your explanation.

Going Further

Try the experiment again, but use ice water instead of room-temperature water. Explain your results in terms of the effects of temperature.

A Sugar Cube Race!

If you drop a sugar cube into a glass of water, how long will it take to dissolve? Will it take 5 minutes, 10 minutes, or longer? What can you do to speed up the rate at which it dissolves? Should you change something about the water, the sugar cube, or the process? In other words, what variable should you change? Before reading further, make a list of variables that could be changed in this situation. Record your list in your ScienceLog.

Materials

- water
- graduated cylinder
- 2 sugar cubes
- 2 beakers or other clear containers
- clock or stopwatch
- other materials approved by your teacher

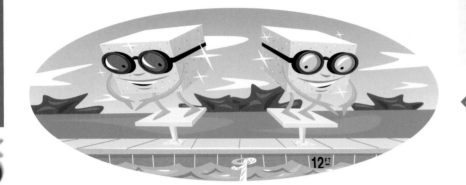

Make a Prediction

1. Choose one variable to test. In your ScienceLog, record your choice, and predict how changing your variable will affect the rate of dissolving.

Conduct an Experiment

2. Pour 150 mL of water into one of the beakers. Add one sugar cube, and use the stopwatch to measure how long it takes for the sugar cube to dissolve. You must not disturb the sugar cube in any way! Record this time in your ScienceLog.

3. Tell your teacher how you wish to test the variable. Do not proceed without his or her approval. You may need additional equipment.

4. Prepare your materials to test the variable you have picked. When you are ready, start your procedure for speeding up the dissolving of the sugar cube. Use the stopwatch to measure the time. Record this time in your ScienceLog.

Analyze the Results

5. Compare your results with the results obtained in step 2. Was your prediction correct? Why or why not?

Draw Conclusions

6. Why was it necessary to observe the sugar cube dissolving on its own before you tested the variable?

7. Do you think that changing more than one variable would speed up the rate of dissolving even more? Explain your reasoning.

Communicate Results

8. Discuss your results with a group that tested a different variable. Which variable had a greater effect on the rate of dissolving?

Making Butter

A colloid is an interesting substance. It has properties of both solutions and suspensions. Colloidal particles are not heavy enough to settle out, so they remain evenly dispersed throughout the mixture. In this activity, you will make butter—a very familiar colloid—and observe the characteristics that classify butter as a colloid.

Materials

- marble
- small, clear container with lid
- heavy cream
- clock or stopwatch

Procedure

1. Place a marble inside the container, and fill the container with heavy cream. Put the lid tightly on the container.

2. Take turns shaking the container vigorously and constantly for 10 minutes. Record the time when you begin shaking in your ScienceLog. Every minute, stop shaking the container and hold it up to the light. Record your observations.

3. Continue shaking the container, taking turns if necessary. When you see, hear, or feel any changes inside the container, note the time and change in your ScienceLog.

4. After 10 minutes of shaking, you should have a lump of "butter" surrounded by liquid inside the container. Describe both the butter and the liquid in detail in your ScienceLog.

5. Let the container sit for about 10 minutes. Observe the butter and liquid again, and record your observations in your ScienceLog.

Analysis

6. When you noticed the change in the container, what did you think was happening at that point?

7. Based on your observations, explain why butter is classified as a colloid.

8. What kind of mixture is the liquid that is left behind? Explain.

141

Unpolluting Water

MAKING MODELS

In many cities, the water supply comes from a river, lake, or reservoir. This water may include several mixtures, including suspensions (with suspended dirt, oil, or living organisms) and solutions (with dissolved chemicals). To make the water safe to drink, your city's water supplier must remove impurities. In this lab, you will model the procedures used in real water-treatment plants.

Part A: Untreated Water

Procedure

1. Measure 100 mL of "polluted" water into a graduated cylinder. Be sure to shake the bottle of water before you pour so your sample will include all the impurities.

2. Pour the contents of the graduated cylinder into one of the beakers.

3. Copy the table below into your ScienceLog, and record your observations of the water in the "Before treatment" row.

Materials

- "polluted" water
- graduated cylinder
- 250 mL beakers (4)
- 2 plastic spoons
- small nail
- 8 oz plastic-foam cup (2)
- scissors
- 2 pieces of filter paper
- washed fine sand
- metric ruler
- washed activated charcoal
- rubber band

Observations						
	Color	**Clearness**	**Odor**	**Any layers?**	**Any solids?**	**Water volume**
Before treatment						
After oil separation						
After sand filtration						
After charcoal						

DO NOT WRITE IN BOOK

Part B: Settling In

If a suspension is left standing, the suspended particles will settle to the top or bottom. You should see a layer of oil at the top.

Procedure

4. Separate the oil by carefully pouring the oil into another beaker. You can use a plastic spoon to get the last bit of oil from the water. Record your observations.

Part C: Filtration

Cloudy water can be a sign of small particles still in suspension. These particles can usually be removed by filtering. Water-treatment plants use sand and gravel as filters.

Procedure

5. Make a filter as follows:
   a. Use the nail to poke 5 to 10 small holes in the bottom of one of the cups.
   b. Cut a circle of filter paper to fit inside the bottom of the cup. (This will keep the sand in the cup.)
   c. Fill the cup to 2 cm below the rim with wet sand. Pack the sand tightly.
   d. Set the cup inside an empty beaker.

6. Pour the polluted water on top of the sand, and let it filter through. Do not pour any of the settled mud onto the sand. (Dispose of the mud as instructed by your teacher.) In your table, record your observations of the water collected in the beaker.

Part D: Separating Solutions

Something that has been dissolved in a solvent cannot be separated using filters. Water-treatment plants use activated charcoal to absorb many dissolved chemicals.

Procedure

7. Place activated charcoal about 3 cm deep in the unused cup. Pour the water collected from the sand filtration into the cup, and stir for a minute with a spoon.

8. Place a piece of filter paper over the top of the cup, and fasten it in place with a rubber band. With the paper securely in place, pour the water through the filter paper and back into a clean beaker. Record your observations in your table.

Analysis (Parts A–D)

9. Is your unpolluted water safe to drink? Why or why not?

10. When you treat a sample of water, do you get out exactly the same amount of water that you put in? Explain your answer.

11. Some groups may still have cloudy water when they finish. Explain a possible cause for this.

Concept Mapping: A Way to Bring Ideas Together

What Is a Concept Map?

Have you ever tried to tell someone about a book or a chapter you've just read and found that you can remember only a few isolated words and ideas? Or maybe you've memorized facts for a test and then weeks later discovered you're not even sure what topics those facts covered.

In both cases, you may have understood the ideas or concepts by themselves but not in relation to one another. If you could somehow link the ideas together, you would probably understand them better and remember them longer. This is something a concept map can help you do. A concept map is a way to see how ideas or concepts fit together. It can help you see the "big picture."

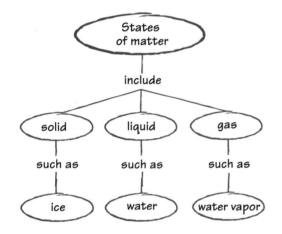

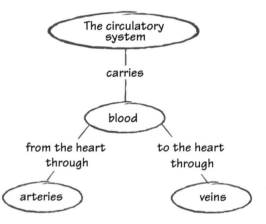

How to Make a Concept Map

1 **Make a list of the main ideas or concepts.**

It might help to write each concept on its own slip of paper. This will make it easier to rearrange the concepts as many times as necessary to make sense of how the concepts are connected. After you've made a few concept maps this way, you can go directly from writing your list to actually making the map.

2 **Arrange the concepts in order from the most general to the most specific.**

Put the most general concept at the top and circle it. Ask yourself, "How does this concept relate to the remaining concepts?" As you see the relationships, arrange the concepts in order from general to specific.

3 **Connect the related concepts with lines.**

4 **On each line, write an action word or short phrase that shows how the concepts are related.**

Look at the concept maps on this page, and then see if you can make one for the following terms:

plants, water, photosynthesis, carbon dioxide, sun's energy

One possible answer is provided at right, but don't look at it until you try the concept map yourself.

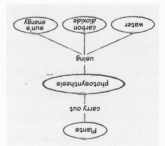

SI Measurement

The International System of Units, or SI, is the standard system of measurement used by many scientists. Using the same standards of measurement makes it easier for scientists to communicate with one another.

SI works by combining prefixes and base units. Each base unit can be used with different prefixes to define smaller and larger quantities. The table below lists common SI prefixes.

SI Prefixes			
Prefix	**Abbreviation**	**Factor**	**Example**
kilo-	k	1,000	kilogram, 1 kg = 1,000 g
hecto-	h	100	hectoliter, 1 hL = 100 L
deka-	da	10	dekameter, 1 dam = 10 m
		1	meter, liter
deci-	d	0.1	decigram, 1 dg = 0.1 g
centi-	c	0.01	centimeter, 1 cm = 0.01 m
milli-	m	0.001	milliliter, 1 mL = 0.001 L
micro-	μ	0.000 001	micrometer, 1 μm = 0.000 001 m

SI Conversion Table		
SI units	**From SI to English**	**From English to SI**
Length		
kilometer (km) = 1,000 m	1 km = 0.621 mi	1 mi = 1.609 km
meter (m) = 100 cm	1 m = 3.281 ft	1 ft = 0.305 m
centimeter (cm) = 0.01 m	1 cm = 0.394 in.	1 in. = 2.540 cm
millimeter (mm) = 0.001 m	1 mm = 0.039 in.	
micrometer (μm) = 0.000 001 m		
nanometer (nm) = 0.000 000 001 m		
Area		
square kilometer (km^2) = 100 hectares	1 km^2 = 0.386 mi^2	1 mi^2 = 2.590 km^2
hectare (ha) = 10,000 m^2	1 ha = 2.471 acres	1 acre = 0.405 ha
square meter (m^2) = 10,000 cm^2	1 m^2 = 10.765 ft^2	1 ft^2 = 0.093 m^2
square centimeter (cm^2) = 100 mm^2	1 cm^2 = 0.155 in.^{2}	1 in.^{2} = 6.452 cm^2
Volume		
liter (L) = 1,000 mL = 1 dm^3	1 L = 1.057 fl qt	1 fl qt = 0.946 L
milliliter (mL) = 0.001 L = 1 cm^3	1 mL = 0.034 fl oz	1 fl oz = 29.575 mL
microliter (μL) = 0.000 001 L		
Mass		
kilogram (kg) = 1,000 g	1 kg = 2.205 lb	1 lb = 0.454 kg
gram (g) = 1,000 mg	1 g = 0.035 oz	1 oz = 28.349 g
milligram (mg) = 0.001 g		
microgram (μg) = 0.000 001 g		

Temperature Scales

Temperature can be expressed using three different scales: Fahrenheit, Celsius, and Kelvin. The SI unit for temperature is the kelvin (K).

Although 0 K is much colder than 0°C, a change of 1 K is equal to a change of 1°C.

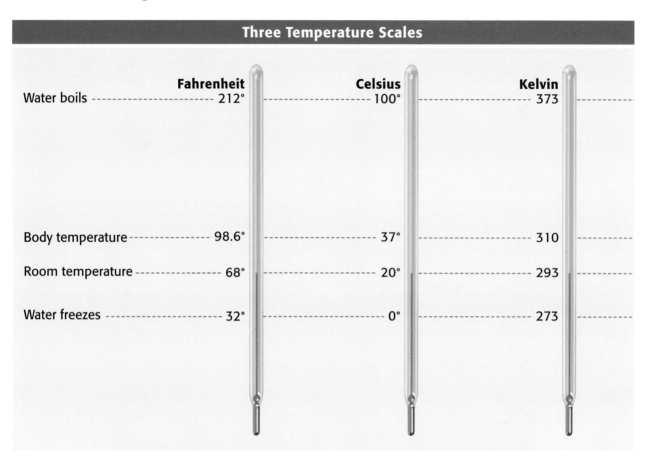

Three Temperature Scales

	Fahrenheit	Celsius	Kelvin
Water boils	212°	100°	373
Body temperature	98.6°	37°	310
Room temperature	68°	20°	293
Water freezes	32°	0°	273

Temperature Conversions Table

To convert	Use this equation:	Example
Celsius to Fahrenheit °C ⟶ °F	$°F = \left(\dfrac{9}{5} \times °C\right) + 32$	Convert 45°C to °F. $°F = \left(\dfrac{9}{5} \times 45°C\right) + 32 = 113°F$
Fahrenheit to Celsius °F ⟶ °C	$°C = \dfrac{5}{9} \times (°F - 32)$	Convert 68°F to °C. $°C = \dfrac{5}{9} \times (68°F - 32) = 20°C$
Celsius to Kelvin °C ⟶ K	$K = °C + 273$	Convert 45°C to K. $K = 45°C + 273 = 318\ K$
Kelvin to Celsius K ⟶ °C	$°C = K - 273$	Convert 32 K to °C. $°C = 32\ K - 273 = -241°C$

Measuring Skills

Using a Graduated Cylinder

When using a graduated cylinder to measure volume, keep the following procedures in mind:

❶ Make sure the cylinder is on a flat, level surface.

❷ Move your head so that your eye is level with the surface of the liquid.

❸ Read the mark closest to the liquid level. On glass graduated cylinders, read the mark closest to the center of the curve in the liquid's surface.

Using a Meterstick or Metric Ruler

When using a meterstick or metric ruler to measure length, keep the following procedures in mind:

❶ Place the ruler firmly against the object you are measuring.

❷ Align one edge of the object exactly with the zero end of the ruler.

❸ Look at the other edge of the object to see which of the marks on the ruler is closest to that edge. **Note:** Each small slash between the centimeters represents a millimeter, which is one-tenth of a centimeter.

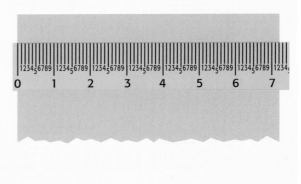

Using a Triple-Beam Balance

When using a triple-beam balance to measure mass, keep the following procedures in mind:

❶ Make sure the balance is on a level surface.

❷ Place all of the countermasses at zero. Adjust the balancing knob until the pointer rests at zero.

❸ Place the object you wish to measure on the pan. **Caution:** Do not place hot objects or chemicals directly on the balance pan.

❹ Move the largest countermass along the beam to the right until it is at the last notch that does not tip the balance. Follow the same procedure with the next-largest countermass. Then move the smallest countermass until the pointer rests at zero.

❺ Add the readings from the three beams together to determine the mass of the object.

❻ When determining the mass of crystals or powders, use a piece of filter paper. First find the mass of the paper. Then add the crystals or powder to the paper and re-measure. The actual mass of the crystals or powder is the total mass minus the mass of the paper. When finding the mass of liquids, first find the mass of the empty container. Then find the mass of the liquid and container together. The mass of the liquid is the total mass minus the mass of the container.

Scientific Method

The series of steps that scientists use to answer questions and solve problems is often called the **scientific method.** The scientific method is not a rigid procedure. Scientists may use all of the steps or just some of the steps of the scientific method. They may even repeat some of the steps. The goal of the scientific method is to come up with reliable answers and solutions.

Six Steps of the Scientific Method

1 **Ask a Question** Good questions come from careful **observations.** You make observations by using your senses to gather information. Sometimes you may use instruments, such as microscopes and telescopes, to extend the range of your senses. As you observe the natural world, you will discover that you have many more questions than answers. These questions drive the scientific method.

Questions beginning with *what, why, how,* and *when* are very important in focusing an investigation, and they often lead to a hypothesis. (You will learn what a hypothesis is in the next step.) Here is an example of a question that could lead to further investigation.

Ask a Question

Question: How does acid rain affect plant growth?

2 **Form a Hypothesis** After you come up with a question, you need to turn the question into a **hypothesis.** A hypothesis is a clear statement of what you expect the answer to your question to be. Your hypothesis will represent your best "educated guess" based on your observations and what you already know. A good hypothesis is testable. If observations and information cannot be gathered or if an experiment cannot be designed to test your hypothesis, it is untestable, and the investigation can go no further.

Form a Hypothesis

Here is a hypothesis that could be formed from the question, "How does acid rain affect plant growth?"

Hypothesis: Acid rain causes plants to grow more slowly.

Notice that the hypothesis provides some specifics that lead to methods of testing. The hypothesis can also lead to predictions. A **prediction** is what you think will be the outcome of your experiment or data collection. Predictions are usually stated in an "if . . . then" format. For example, **if** meat is kept at room temperature, **then** it will spoil faster than meat kept in the refrigerator. More than one prediction can be made for a single hypothesis. Here is a sample prediction for the hypothesis that acid rain causes plants to grow more slowly.

Prediction: If a plant is watered with only acid rain (which has a pH of 4), then the plant will grow at half its normal rate.

3 **Test the Hypothesis** After you have formed a hypothesis and made a prediction, you should test your hypothesis. There are different ways to do this. Perhaps the most familiar way is to conduct a **controlled experiment.** A controlled experiment tests only one factor at a time. A controlled experiment has a **control group** and one or more **experimental groups.** All the factors for the control and experimental groups are the same except for one factor, which is called the **variable.** By changing only one factor, you can see the results of just that one change.

Sometimes, the nature of an investigation makes a controlled experiment impossible. For example, dinosaurs have been extinct for millions of years, and the Earth's core is surrounded by thousands of meters of rock. It would be difficult, if not impossible, to conduct controlled experiments on such things. Under such circumstances, a hypothesis may be tested by making detailed observations. Taking measurements is one way of making observations.

Test the Hypothesis

4 **Analyze the Results** After you have completed your experiments, made your observations, and collected your data, you must analyze all the information you have gathered. Tables and graphs are often used in this step to organize the data.

Analyze the Results

5 **Draw Conclusions** Based on the analysis of your data, you should conclude whether or not your results support your hypothesis. If your hypothesis is supported, you (or others) might want to repeat the observations or experiments to verify your results. If your hypothesis is not supported by the data, you may have to check your procedure for errors. You may even have to reject your hypothesis and make a new one. If you cannot draw a conclusion from your results, you may have to try the investigation again or carry out further observations or experiments.

Draw Conclusions

Do they support your hypothesis?

No

Yes

6 **Communicate Results** After any scientific investigation, you should report your results. By doing a written or oral report, you let others know what you have learned. They may want to repeat your investigation to see if they get the same results. Your report may even lead to another question, which in turn may lead to another investigation.

Communicate Results

APPENDIX

Scientific Method in Action

The scientific method is not a "straight line" of steps. It contains loops in which several steps may be repeated over and over again, while others may not be necessary. For example, sometimes scientists will find that testing one hypothesis raises new questions and new hypotheses to be tested. And sometimes, testing the hypothesis leads directly to a conclusion. Furthermore, the steps in the scientific method are not always used in the same order. Follow the steps in the diagram below, and see how many different directions the scientific method can take you.

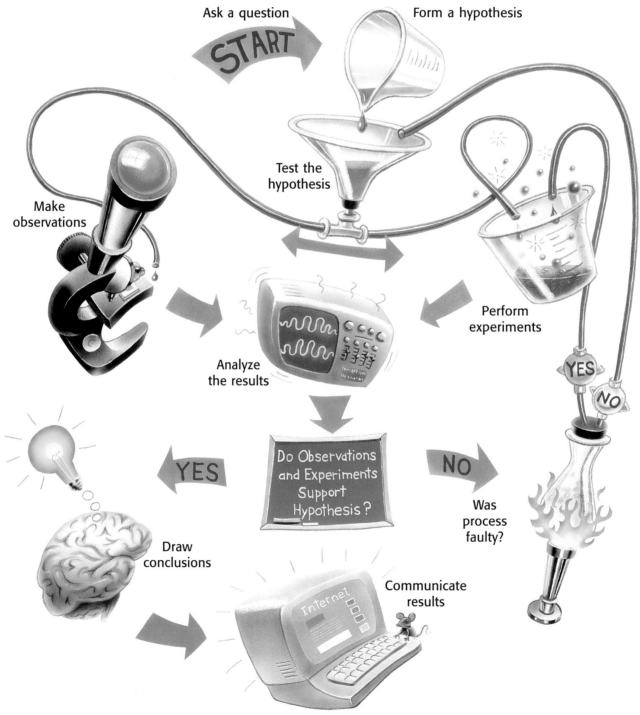

Making Charts and Graphs

Circle Graphs

A circle graph, or pie chart, shows how each group of data relates to all of the data. Each part of the circle represents a category of the data. The entire circle represents all of the data. For example, a biologist studying a hardwood forest in Wisconsin found that there were five different types of trees. The data table at right summarizes the biologist's findings.

Wisconsin Hardwood Trees	
Type of tree	**Number found**
Oak	600
Maple	750
Beech	300
Birch	1,200
Hickory	150
Total	3,000

How to Make a Circle Graph

1 In order to make a circle graph of this data, first find the percentage of each type of tree. To do this, divide the number of individual trees by the total number of trees and multiply by 100.

$$\frac{600 \text{ oak}}{3{,}000 \text{ trees}} \times 100 = 20\%$$

$$\frac{750 \text{ maple}}{3{,}000 \text{ trees}} \times 100 = 25\%$$

$$\frac{300 \text{ beech}}{3{,}000 \text{ trees}} \times 100 = 10\%$$

$$\frac{1{,}200 \text{ birch}}{3{,}000 \text{ trees}} \times 100 = 40\%$$

$$\frac{150 \text{ hickory}}{3{,}000 \text{ trees}} \times 100 = 5\%$$

2 Now determine the size of the pie shapes that make up the chart. Do this by multiplying each percentage by 360°. Remember that a circle contains 360°.

$20\% \times 360° = 72°$ $25\% \times 360° = 90°$
$10\% \times 360° = 36°$ $40\% \times 360° = 144°$
$5\% \times 360° = 18°$

3 Then check that the sum of the percentages is 100 and the sum of the degrees is 360.

$20\% + 25\% + 10\% + 40\% + 5\% = 100\%$
$72° + 90° + 36° + 144° + 18° = 360°$

4 Use a compass to draw a circle and mark its center.

5 Then use a protractor to draw angles of 72°, 90°, 36°, 144°, and 18° in the circle.

6 Finally, label each part of the graph, and choose an appropriate title.

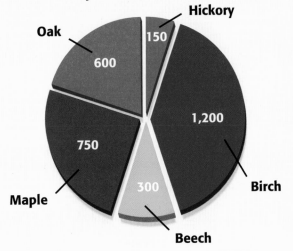

A Community of Wisconsin Hardwood Trees

Hickory 150
Oak 600
Birch 1,200
Maple 750
Beech 300

Population of Appleton, 1900–2000	
Year	Population
1900	1,800
1920	2,500
1940	3,200
1960	3,900
1980	4,600
2000	5,300

Line Graphs

Line graphs are most often used to demonstrate continuous change. For example, Mr. Smith's science class analyzed the population records for their hometown, Appleton, between 1900 and 2000. Examine the data at left.

Because the year and the population change, they are the *variables*. The population is determined by, or dependent on, the year. Therefore, the population is called the **dependent variable**, and the year is called the **independent variable**. Each set of data is called a **data pair.** To prepare a line graph, data pairs must first be organized in a table like the one at left.

How to Make a Line Graph

❶ Place the independent variable along the horizontal (*x*) axis. Place the dependent variable along the vertical (*y*) axis.

❷ Label the *x*-axis "Year" and the *y*-axis "Population." Look at your largest and smallest values for the population. Determine a scale for the *y*-axis that will provide enough space to show these values. You must use the same scale for the entire length of the axis. Find an appropriate scale for the *x*-axis too.

❸ Choose reasonable starting points for each axis.

❹ Plot the data pairs as accurately as possible.

❺ Choose a title that accurately represents the data.

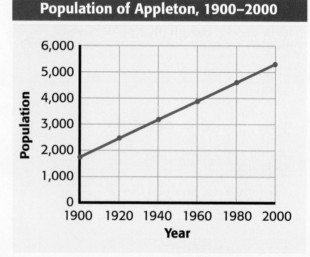

Population of Appleton, 1900–2000

How to Determine Slope

Slope is the ratio of the change in the *y*-axis to the change in the *x*-axis, or "rise over run."

❶ Choose two points on the line graph. For example, the population of Appleton in 2000 was 5,300 people. Therefore, you can define point *a* as (2000, 5,300). In 1900, the population was 1,800 people. Define point *b* as (1900, 1,800).

❷ Find the change in the *y*-axis.
(*y* at point *a*) − (*y* at point *b*)
5,300 people − 1,800 people = 3,500 people

❸ Find the change in the *x*-axis.
(*x* at point *a*) − (*x* at point *b*)
2000 − 1900 = 100 years

❹ Calculate the slope of the graph by dividing the change in *y* by the change in *x*.

$$\text{slope} = \frac{\text{change in } y}{\text{change in } x}$$

$$\text{slope} = \frac{3{,}500 \text{ people}}{100 \text{ years}}$$

slope = 35 people per year

In this example, the population in Appleton increased by a fixed amount each year. The graph of this data is a straight line. Therefore, the relationship is **linear.** When the graph of a set of data is not a straight line, the relationship is **nonlinear.**

Using Algebra to Determine Slope

The equation in step 4 may also be arranged to be:

$$y = kx$$

where y represents the change in the y-axis, k represents the slope, and x represents the change in the x-axis.

$$\text{slope} = \frac{\text{change in } y}{\text{change in } x}$$

$$k = \frac{y}{x}$$

$$k \times x = \frac{y \times x}{x}$$

$$kx = y$$

Bar Graphs

Bar graphs are used to demonstrate change that is not continuous. These graphs can be used to indicate trends when the data are taken over a long period of time. A meteorologist gathered the precipitation records at right for Hartford, Connecticut, for April 1–15, 1996, and used a bar graph to represent the data.

Precipitation in Hartford, Connecticut April 1–15, 1996

Date	Precipitation (cm)	Date	Precipitation (cm)
April 1	0.5	April 9	0.25
April 2	1.25	April 10	0.0
April 3	0.0	April 11	1.0
April 4	0.0	April 12	0.0
April 5	0.0	April 13	0.25
April 6	0.0	April 14	0.0
April 7	0.0	April 15	6.50
April 8	1.75		

How to Make a Bar Graph

1. Use an appropriate scale and a reasonable starting point for each axis.
2. Label the axes, and plot the data.
3. Choose a title that accurately represents the data.

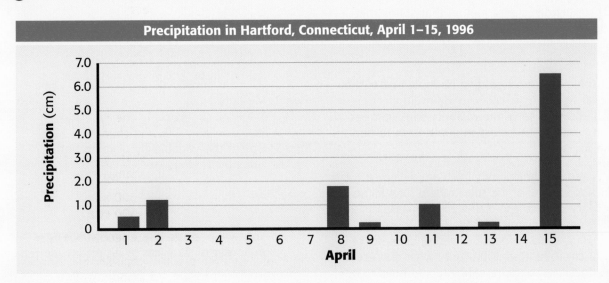

Math Refresher

Science requires an understanding of many math concepts. The following pages will help you review some important math skills.

Averages

An **average,** or **mean,** simplifies a list of numbers into a single number that *approximates* their value.

> **Example:** Find the average of the following set of numbers: 5, 4, 7, and 8.

Step 1: Find the sum.

$$5 + 4 + 7 + 8 = 24$$

Step 2: Divide the sum by the amount of numbers in your set. Because there are four numbers in this example, divide the sum by 4.

$$\frac{24}{4} = 6$$

The average, or mean, is **6.**

Ratios

A **ratio** is a comparison between numbers, and it is usually written as a fraction.

> **Example:** Find the ratio of thermometers to students if you have 36 thermometers and 48 students in your class.

Step 1: Make the ratio.

$$\frac{36 \text{ thermometers}}{48 \text{ students}}$$

Step 2: Reduce the fraction to its simplest form.

$$\frac{36}{48} = \frac{36 \div 12}{48 \div 12} = \frac{3}{4}$$

The ratio of thermometers to students is **3 to 4,** or $\frac{3}{4}$. The ratio may also be written in the form 3:4.

Proportions

A **proportion** is an equation that states that two ratios are equal.

$$\frac{3}{1} = \frac{12}{4}$$

To solve a proportion, first multiply across the equal sign. This is called cross-multiplication. If you know three of the quantities in a proportion, you can use cross-multiplication to find the fourth.

> **Example:** Imagine that you are making a scale model of the solar system for your science project. The diameter of Jupiter is 11.2 times the diameter of the Earth. If you are using a plastic-foam ball with a diameter of 2 cm to represent the Earth, what diameter does the ball representing Jupiter need to be?
>
> $$\frac{11.2}{1} = \frac{x}{2 \text{ cm}}$$

Step 1: Cross-multiply.

$$\frac{11.2}{1} \diagdown\diagup \frac{x}{2}$$

$$11.2 \times 2 = x \times 1$$

Step 2: Multiply.

$$22.4 = x \times 1$$

Step 3: Isolate the variable by dividing both sides by 1.

$$x = \frac{22.4}{1}$$
$$x = 22.4 \text{ cm}$$

You will need to use a ball with a diameter of **22.4 cm** to represent Jupiter.

Percentages

A **percentage** is a ratio of a given number to 100.

> **Example:** What is 85 percent of 40?

Step 1: Rewrite the percentage by moving the decimal point two places to the left.

$$.85$$

Step 2: Multiply the decimal by the number you are calculating the percentage of.

$$0.85 \times 40 = 34$$

85 percent of 40 is **34.**

Decimals

To **add** or **subtract decimals,** line up the digits vertically so that the decimal points line up. Then add or subtract the columns from right to left, carrying or borrowing numbers as necessary.

> **Example:** Add the following numbers: 3.1415 and 2.96.

Step 1: Line up the digits vertically so that the decimal points line up.

$$
\begin{array}{r}
3.1415 \\
+\ 2.96 \\
\hline
\end{array}
$$

Step 2: Add the columns from right to left, carrying when necessary.

$$
\begin{array}{r}
1\ 1 \\
3.1415 \\
+\ 2.96 \\
\hline
6.1015
\end{array}
$$

The sum is **6.1015.**

Fractions

Numbers tell you how many; **fractions** tell you *how much of a whole.*

> **Example:** Your class has 24 plants. Your teacher instructs you to put 5 in a shady spot. What fraction does this represent?

Step 1: Write a fraction with the total number of parts in the whole as the denominator.

$$\frac{?}{24}$$

Step 2: Write the number of parts of the whole being represented as the numerator.

$$\frac{5}{24}$$

$\frac{5}{24}$ of the plants will be in the shade.

Reducing Fractions

It is usually best to express a fraction in simplest form. This is called *reducing* a fraction.

> **Example:** Reduce the fraction $\frac{30}{45}$ to its simplest form.

Step 1: Find the largest whole number that will divide evenly into both the numerator and denominator. This number is called the greatest common factor (GCF).

factors of the numerator 30: 1, 2, 3, 5, 6, 10, **15,** 30

factors of the denominator 45: 1, 3, 5, 9, **15,** 45

Step 2: Divide both the numerator and the denominator by the GCF, which in this case is 15.

$$\frac{30}{45} = \frac{30 \div 15}{45 \div 15} = \frac{2}{3}$$

$\frac{30}{45}$ reduced to its simplest form is $\frac{2}{3}$.

APPENDIX

Adding and Subtracting Fractions

To **add** or **subtract fractions** that have the **same denominator,** simply add or subtract the numerators.

> **Examples:**
> $$\frac{3}{5} + \frac{1}{5} = ? \quad \text{and} \quad \frac{3}{4} - \frac{1}{4} = ?$$

Step 1: Add or subtract the numerators.
$$\frac{3}{5} + \frac{1}{5} = \frac{4}{\quad} \quad \text{and} \quad \frac{3}{4} - \frac{1}{4} = \frac{2}{\quad}$$

Step 2: Write the sum or difference over the denominator.
$$\frac{3}{5} + \frac{1}{5} = \frac{4}{5} \quad \text{and} \quad \frac{3}{4} - \frac{1}{4} = \frac{2}{4}$$

Step 3: If necessary, reduce the fraction to its simplest form.
$$\frac{4}{5} \text{ cannot be reduced, and } \frac{2}{4} = \frac{1}{2}.$$

To **add** or **subtract fractions** that have **different denominators,** first find the least common denominator (LCD).

> **Examples:**
> $$\frac{1}{2} + \frac{1}{6} = ? \quad \text{and} \quad \frac{3}{4} - \frac{2}{3} = ?$$

Step 1: Write the equivalent fractions with a common denominator.
$$\frac{3}{6} + \frac{1}{6} = ? \quad \text{and} \quad \frac{9}{12} - \frac{8}{12} = ?$$

Step 2: Add or subtract.
$$\frac{3}{6} + \frac{1}{6} = \frac{4}{6} \quad \text{and} \quad \frac{9}{12} - \frac{8}{12} = \frac{1}{12}$$

Step 3: If necessary, reduce the fraction to its simplest form.
$$\frac{4}{6} = \frac{2}{3}, \text{ and } \frac{1}{12} \text{ cannot be reduced.}$$

Multiplying Fractions

To **multiply fractions,** multiply the numerators and the denominators together, and then reduce the fraction to its simplest form.

> **Example:**
> $$\frac{5}{9} \times \frac{7}{10} = ?$$

Step 1: Multiply the numerators and denominators.
$$\frac{5}{9} \times \frac{7}{10} = \frac{5 \times 7}{9 \times 10} = \frac{35}{90}$$

Step 2: Reduce.
$$\frac{35}{90} = \frac{35 \div 5}{90 \div 5} = \frac{7}{18}$$

Dividing Fractions

To **divide fractions,** first rewrite the divisor (the number you divide *by*) upside down. This is called the reciprocal of the divisor. Then you can multiply and reduce if necessary.

> **Example:**
> $$\frac{5}{8} \div \frac{3}{2} = ?$$

Step 1: Rewrite the divisor as its reciprocal.
$$\frac{3}{2} \rightarrow \frac{2}{3}$$

Step 2: Multiply.
$$\frac{5}{8} \times \frac{2}{3} = \frac{5 \times 2}{8 \times 3} = \frac{10}{24}$$

Step 3: Reduce.
$$\frac{10}{24} = \frac{10 \div 2}{24 \div 2} = \frac{5}{12}$$

156   Appendix

Scientific Notation

Scientific notation is a short way of representing very large and very small numbers without writing all of the place-holding zeros.

Example: Write 653,000,000 in scientific notation.

Step 1: Write the number without the place-holding zeros.

653

Step 2: Place the decimal point after the first digit.

6.53

Step 3: Find the exponent by counting the number of places that you moved the decimal point.

6.53000000

The decimal point was moved eight places to the left. Therefore, the exponent of 10 is positive 8. Remember, if the decimal point had moved to the right, the exponent would be negative.

Step 4: Write the number in scientific notation.

$$6.53 \times 10^8$$

Area

Area is the number of square units needed to cover the surface of an object.

Formulas:
Area of a square = side × side
Area of a rectangle = length × width
Area of a triangle = $\frac{1}{2}$ × base × height

Examples: Find the areas.

Triangle
Area = $\frac{1}{2}$ × base × height
Area = $\frac{1}{2}$ × 3 cm × 4 cm
Area = **6 cm²**

Rectangle
Area = length × width
Area = 6 cm × 3 cm
Area = **18 cm²**

Square
Area = side × side
Area = 3 cm × 3 cm
Area = **9 cm²**

Volume

Volume is the amount of space something occupies.

Formulas:
Volume of a cube =
side × side × side

Volume of a prism =
area of base × height

Examples:
Find the volume
of the solids.

Cube
Volume = side × side × side
Volume = 4 cm × 4 cm × 4 cm
Volume = **64 cm³**

Prism
Volume = area of base × height
Volume = (area of triangle) × height
Volume = $\left(\frac{1}{2} \times 3 \text{ cm} \times 4 \text{ cm} \right) \times 5$ cm
Volume = 6 cm² × 5 cm
Volume = **30 cm³**

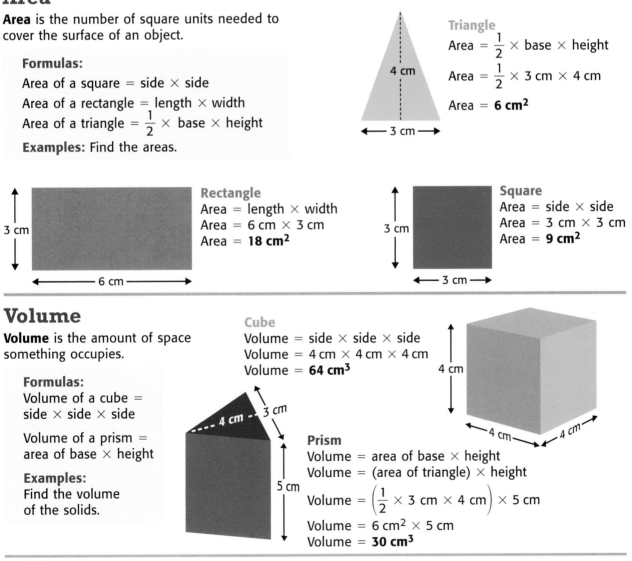

Periodic Table of the Elements

Each square on the table includes an element's name, chemical symbol, atomic number, and atomic mass.

Atomic number ———— 6
Chemical symbol ———— C
Element name ———— Carbon
Atomic mass ———— 12.0

The background color indicates the type of element. Carbon is a nonmetal.

The color of the chemical symbol indicates the physical state at room temperature. Carbon is a solid.

Background
Metals
Metalloids
Nonmetals

Chemical Symbol
Solid
Liquid
Gas

Period 1
1
H
Hydrogen
1.0

Group 1	Group 2
3 Li Lithium 6.9	4 Be Beryllium 9.0
11 Na Sodium 23.0	12 Mg Magnesium 24.3

		Group 3	Group 4	Group 5	Group 6	Group 7	Group 8	Group 9
Period 4 19 K Potassium 39.1	20 Ca Calcium 40.1	21 Sc Scandium 45.0	22 Ti Titanium 47.9	23 V Vanadium 50.9	24 Cr Chromium 52.0	25 Mn Manganese 54.9	26 Fe Iron 55.8	27 Co Cobalt 58.9
Period 5 37 Rb Rubidium 85.5	38 Sr Strontium 87.6	39 Y Yttrium 88.9	40 Zr Zirconium 91.2	41 Nb Niobium 92.9	42 Mo Molybdenum 95.9	43 Tc Technetium (97.9)	44 Ru Ruthenium 101.1	45 Rh Rhodium 102.9
Period 6 55 Cs Cesium 132.9	56 Ba Barium 137.3	57 La Lanthanum 138.9	72 Hf Hafnium 178.5	73 Ta Tantalum 180.9	74 W Tungsten 183.8	75 Re Rhenium 186.2	76 Os Osmium 190.2	77 Ir Iridium 192.2
Period 7 87 Fr Francium (223.0)	88 Ra Radium (226.0)	89 Ac Actinium (227.0)	104 Rf Rutherfordium (261.1)	105 Db Dubnium (262.1)	106 Sg Seaborgium (263.1)	107 Bh Bohrium (262.1)	108 Hs Hassium (265)	109 Mt Meitnerium (266)

A row of elements is called a period.

A column of elements is called a group or family.

Lanthanides

| 58 Ce Cerium 140.1 | 59 Pr Praseodymium 140.9 | 60 Nd Neodymium 144.2 | 61 Pm Promethium (144.9) | 62 Sm Samarium 150.4 |

Actinides

| 90 Th Thorium 232.0 | 91 Pa Protactinium 231.0 | 92 U Uranium 238.0 | 93 Np Neptunium (237.0) | 94 Pu Plutonium 244.1 |

These elements are placed below the table to allow the table to be narrower.

This zigzag line reminds you where the metals, nonmetals, and metalloids are.

Group 18

| 2 **He** Helium 4.0 |

Group 13	**Group 14**	**Group 15**	**Group 16**	**Group 17**
5 **B** Boron 10.8	6 **C** Carbon 12.0	7 **N** Nitrogen 14.0	8 **O** Oxygen 16.0	9 **F** Fluorine 19.0	10 **Ne** Neon 20.2

| | | | 13 **Al** Aluminum 27.0 | 14 **Si** Silicon 28.1 | 15 **P** Phosphorus 31.0 | 16 **S** Sulfur 32.1 | 17 **Cl** Chlorine 35.5 | 18 **Ar** Argon 39.9 |

Group 10	**Group 11**	**Group 12**						
28 **Ni** Nickel 58.7	29 **Cu** Copper 63.5	30 **Zn** Zinc 65.4	31 **Ga** Gallium 69.7	32 **Ge** Germanium 72.6	33 **As** Arsenic 74.9	34 **Se** Selenium 79.0	35 **Br** Bromine 79.9	36 **Kr** Krypton 83.8
46 **Pd** Palladium 106.4	47 **Ag** Silver 107.9	48 **Cd** Cadmium 112.4	49 **In** Indium 114.8	50 **Sn** Tin 118.7	51 **Sb** Antimony 121.8	52 **Te** Tellurium 127.6	53 **I** Iodine 126.9	54 **Xe** Xenon 131.3
78 **Pt** Platinum 195.1	79 **Au** Gold 197.0	80 **Hg** Mercury 200.6	81 **Tl** Thallium 204.4	82 **Pb** Lead 207.2	83 **Bi** Bismuth 209.0	84 **Po** Polonium (209.0)	85 **At** Astatine (210.0)	86 **Rn** Radon (222.0)
110 **Uun*** Ununnilium (271)	111 **Uuu*** Unununium (272)	112 **Uub*** Ununbium (277)		114 **Uuq*** Ununquadium (285)		116 **Uuh*** Ununhexium (289)		118 **Uuo*** Ununoctium (293)

A number in parenthesis is the mass number of the most stable form of that element.

63 **Eu** Europium 152.0	64 **Gd** Gadolinium 157.3	65 **Tb** Terbium 158.9	66 **Dy** Dysprosium 162.5	67 **Ho** Holmium 164.9	68 **Er** Erbium 167.3	69 **Tm** Thulium 168.9	70 **Yb** Ytterbium 173.0	71 **Lu** Lutetium 175.0
95 **Am** Americium (243.1)	96 **Cm** Curium (247.1)	97 **Bk** Berkelium (247.1)	98 **Cf** Californium (251.1)	99 **Es** Einsteinium (252.1)	100 **Fm** Fermium (257.1)	101 **Md** Mendelevium (258.1)	102 **No** Nobelium (259.1)	103 **Lr** Lawrencium (262.1)

*The official names and symbols for the elements greater than 109 will eventually be approved by a committee of scientists.

Glossary

A

alkaline-earth metals the elements in Group 2 of the periodic table; they are reactive metals but are less reactive than alkali metals; their atoms have two electrons in their outer level (113)

alloys solid solutions of metals or nonmetals dissolved in metals (65)

atom the smallest particle into which an element can be divided and still be the same substance (80)

atomic mass the weighted average of the masses of all the naturally occurring isotopes of an element (92)

atomic mass unit (amu) the SI unit used to express the masses of particles in atoms (88)

atomic number the number of protons in the nucleus of an atom (90)

B

boiling vaporization that occurs throughout a liquid (40)

boiling point the temperature at which a liquid boils and becomes a gas (40)

Boyle's law the law that states that for a fixed amount of gas at a constant temperature, the volume of a gas increases as its pressure decreases (35)

C

change of state the conversion of a substance from one physical form to another (38)

characteristic property a property of a substance that is always the same whether the sample observed is large or small (16)

Charles's law the law that states that for a fixed amount of gas at a constant pressure, the volume of a gas increases as its temperature increases (36)

chemical change a change that occurs when one or more substances are changed into entirely new substances with different properties; cannot be reversed using physical means (17)

chemical property a property of matter that describes a substance based on its ability to change into a new substance with different properties (15)

colloid (KAWL oyd) a mixture in which the particles are dispersed throughout but are not heavy enough to settle out (69)

compound a pure substance composed of two or more elements that are chemically combined (58)

concentration a measure of the amount of solute dissolved in a solvent (66)

condensation the change of state from a gas to a liquid (41)

condensation point the temperature at which a gas becomes a liquid (41)

D

density the amount of matter in a given space; mass per unit volume (12)

ductility (duhk TIL uh tee) the ability of a substance to be drawn or pulled into a wire (12)

E

electron clouds the regions inside an atom where electrons are likely to be found (86)

electrons the negatively charged particles found in all atoms; electrons are involved in the formation of chemical bonds (83)

element a pure substance that cannot be separated or broken down into simpler substances by physical or chemical means (54)

endothermic the term used to describe a physical or a chemical change in which energy is absorbed (39)

evaporation (ee VAP uh RAY shuhn) vaporization that occurs at the surface of a liquid below its boiling point (40)

exothermic the term used to describe a physical or a chemical change in which energy is released or removed (39)

F

freezing the change of state from a liquid to a solid (39)

freezing point the temperature at which a liquid changes into a solid (39)

G

gas the state in which matter changes in both shape and volume (33)

gravity a force of attraction between objects that is due to their masses (7)

group a column of elements on the periodic table (111)

H

halogens the elements in Group 17 of the periodic table; they are very reactive nonmetals, and their atoms have seven electrons in their outer level (118)

heterogeneous (HET uhr OH JEE nee uhs) **mixture** a combination of substances in which different components are easily observed (68)

homogeneous (HOH moh JEE nee uhs) **mixture** a combination of substances in which the appearance and properties are the same throughout (64)

hypothesis a possible explanation or answer to a question (148)

I

inertia the tendency of all objects to resist any change in motion (10)

isotopes atoms that have the same number of protons but have different numbers of neutrons (90)

L

liquid the state in which matter takes the shape of its container and has a definite volume (32)

M

malleability (MAL ee uh BIL uh tee) the ability of a substance to be pounded into thin sheets (12)

mass the amount of matter that something is made of (6)

mass number the sum of the protons and neutrons in an atom (91)

matter anything that has volume and mass (4)

melting the change of state from a solid to a liquid (39)

melting point the temperature at which a substance changes from a solid to a liquid (39)

meniscus (muh NIS kuhs) the curve at a liquid's surface by which you measure the volume of the liquid (5)

metalloids elements that have properties of both metals and nonmetals; sometimes referred to as semiconductors (57)

metals elements that are shiny and are good conductors of thermal energy and electric current; most metals are malleable and ductile (57)

mixture a combination of two or more substances that are not chemically combined (62)

model a representation of an object or system (83)

N

neutrons the particles of the nucleus that have no charge (88)

newton (N) the SI unit of force (9)

noble gases the unreactive elements in Group 18 of the periodic table; their atoms have eight electrons in their outer level (except for helium, which has two electrons) (118)

nonmetals elements that are dull and are poor conductors of thermal energy and electric current (57)

nucleus (NOO klee uhs) the tiny, extremely dense, positively charged region in the center of an atom; made up of protons and neutrons (85)

P

period a horizontal row of elements on the periodic table (111)

periodic having a regular, repeating pattern (104)

periodic law the law that states that the chemical and physical properties of elements are periodic functions of their atomic numbers (105)

physical change a change that affects one or more physical properties of a substance; many physical changes are easy to undo (16)

physical property a property of matter that can be observed or measured without changing the identity of the matter (11)

plasma the state of matter that does not have a definite shape or volume and whose particles have broken apart; plasma is composed of electrons and positively charged ions (37)

protons the positively charged particles of the nucleus; the number of protons in a nucleus is the atomic number that determines the identity of an element (88)

S

saturated solution a solution that contains all the solute it can hold at a given temperature (66)

scientific method a series of steps that scientists use to answer questions and solve problems (148)

solid the state in which matter has a definite shape and volume (31)

solubility (SAHL yoo BIL uh tee) the ability to dissolve in another substance; more specifically, the amount of solute needed to make a saturated solution using a given amount of solvent at a certain temperature (12, 66)

solute the substance that is dissolved to form a solution (64)

solution a mixture that appears to be a single substance but is composed of particles of two or more substances that are distributed evenly amongst each other (64)

solvent the substance in which a solute is dissolved to form a solution (64)

states of matter the physical forms in which a substance can exist; states include solid, liquid, gas, and plasma (30)

sublimation (SUHB luh MAY shuhn) the change of state from a solid directly into a gas (42)

surface tension the force acting on the particles at the surface of a liquid that causes the liquid to form spherical drops (33)

suspension a mixture in which particles of a material are dispersed throughout a liquid or gas but are large enough that they settle out (68)

T

theory a unifying explanation for a broad range of hypotheses and observations that have been supported by testing (80)

V

vaporization the change of state from a liquid to a gas; includes boiling and evaporation (40)

viscosity (vis KAHS uh tee) a liquid's resistance to flow (33)

volume the amount of space that something occupies or the amount of space that something contains (4)

W

weight a measure of the gravitational force exerted on an object, usually by the Earth (8)

Index

Credits

Abbreviations used: (t) top, (c) center, (b) bottom, (l) left, (r) right, (bkgd) background

ILLUSTRATIONS

All illustrations, unless noted below, by Holt, Rinehart and Winston.

Table of Contents Page iv(tl), Kristy Sprott; iv(br), Dan Stuckenschneider/Uhl Studios, Inc.

Chapter One Page 6(t), 7, Stephen Durke/Washington Artists; 10(l), Gary Locke/Suzanne Craig; 11, Blake Thornton/Rita Marie; 18, 19, 23, Marty Roper/Planet Rep; 25(lc), Terry Kovalcik; 27(tc), Daniels & Daniels.

Chapter Two Page 30(t), Mark Heine; 30(b), 31, 32, 33, 34, 35, 36(cl,cr), Stephen Durke/Washington Artists; 36(bl), Preface, Inc.; 38, David Schleinkofer/Mendola Ltd.; 40(t), Marty Roper/Planet Rep; 40(b), Mark Heine; 43, David Schleinkofer/Mendola Ltd. and Preface, Inc.; 46(t), Stephen Durke/Washington Artists; 46(b), Preface, Inc.; 47, Marty Roper/Planet Rep; 49(cr), Preface, Inc.

Chapter Three Page 54, Marty Roper/Planet Rep; 56(b), Preface, Inc.; 60, Blake Thornton/Rita Marie; 67, Preface, Inc.

Chapter Four Page 81(c), Preface, Inc.; 82, Mark Heine; 83, Stephen Durke/Washington Artists; 84(c), Mark Heine; 84(b), Preface, Inc.; 85(t), Stephen Durke/Washington Artists; 85(b), Stephen Durke/Washington Artists; 86, 88, 89(t,b), Stephen Durke/Washington Artists; 89(cr), Terry Kovalcik; 90, 91(b), 93, 94(br), Stephen Durke/Washington Artists; 97, Terry Kovalcik; 98, Mark Heine; 99(r), Stephen Durke/Washington Artists.

Chapter Five Page 104, Michael Jaroszko/American Artists; 106, 107, Kristy Sprott; 108(tr), 109, Stephen Durke/Washington Artists; 111, 112(bc), 113(bl), 114(t), 115(t), 116(tc, b), Preface, Inc.; 116(l), Gary Locke/Suzanne Craig; 117, 118, 119(lc), Preface, Inc.; 123, Gary Locke/Suzanne Craig; 125(tr), Preface, Inc.; 125(bl), Keith Locke/Suzanne Craig; 125(br), Annie Bissett; 126-127(l), Dan Stuckenschneider/Uhl Studios Inc.

LabBook Page 140, Blake Thornton/Rita Marie.

Appendix Page 146(t), Terry Guyer; 150(b), Mark Mille/Sharon Langley Artist Rep.; 151, 152, 153, Preface, Inc.; 158, 159, Kristy Sprott.

PHOTOGRAPHY

Front Cover John Higginson/Stone

Table of Contents iv(bl), Victoria Smith/HRW Photo; v(l), Runk/Schoenberg/Grant Heilman Photography, Inc., Richard Megna/Fundamental Photographs; v(r), Joseph Drivas/The Image Bank; vi(tl), Dr. Harold E. Edgerton/©The Harold E. Edgerton 1992 Trust/courtesy Palm Press, Inc.; vi(cl), Stuart Westmoreland/Stone; vi(br), Scott Van Osdol/HRW Photo; vii(tr), Brett H. Froomer/The Image Bank; vii(cr), Kennan Ward Photography; vii(br), Scott Van Osdol/HRW Photo

Feature Borders Unless otherwise noted below, all images ©2001 PhotoDisc/HRW: "Across the Sciences" Pages 26, 100, all images by HRW; "Careers" 101, sand bkgd and saturn, Corbis Images, DNA, Morgan Cain & Associates, scuba gear, ©1997 Radlund & Associates for Artville; "Eureka" 51, ©2001 PhotoDisc/HRW; "Health Watch" 27, dumbbell, Sam Dudgeon/HRW Photo, aloe vera and EKG, Victoria Smith/HRW Photo, basketball, ©1997 Radlund & Associates for Artville, shoes and Bubbles, Greg Geisler; "Science Fiction" 77, saucers, Ian Christopher/Greg Geisler, book, HRW, bkgd, Stock Illustration Source; "Science, Technology, and Society" 50, 76, 126, robot, Greg Geisler; "Weird Science" 127, mite, David Burder/Stone, atom balls, J/B Woolsey Associates, walking stick and turtle, EclectiCollection.

Chapter One pp. 2-3 Ken Reid/FPG International; 3 HRW Photo; 4(b), NASA, Media Services Corp.; 8(all), John Morrison/Morrison Photography; 9(t), Michelle Bridwell/HRW Photo; 11-12(all), John Morrison/Morrison Photography; 13, Neal Nishler/Stone; 14(all), John Morrison/Morrison Photography; 15, Rob Boudreau/Stone; 16(t), Brett H. Froomer/The Image Bank; 16(cl,cr), John Morrison/Morrison Photography; 17, Lance Schriner/HRW Photo; 18(tl,tr), John Morrison/Morrison Photography; 18(bl), Joseph Drivas/The Image Bank; 18(br), SuperStock; 22(r), Michelle Bridwell/HRW Photo; 25, Lance Schriner/HRW Photo; 26, David Malin/©Anglo-Australian Observatory/Royal Observatory, Edinburgh.

Chapter Two pp. 28-29 Phil Degginger/Stone; 29 HRW Photo; 31(t), Gilbert J. Charbonneau; 33(t), Dr. Harold E. Edgerton/©The Harold E. Edgerton 1992 Trust/courtesy Palm Press, Inc.; 37(b), Pekka Parviainen/Science Photo Library/ Photo Researchers, Inc.; 38, Union Pacific Museum Collection; 39(t), Richard Megna/Fundamental Photographs; 44 Victoria Smith/HRW Photo; 45 Victoria Smith/HRW Photo; 48(t), Myrleen Ferguson/PhotoEdit; 49, Charles D. Winters/ Photo Researchers, Inc.; 50(l), Kennan Ward Photography; 50(r), Dr. Jean-Claude Diels/University of New Mexico; 51, Union Pacific Museum Collection.

Chapter Three pp. 52-53 Scott Van Osdol/HRW Photo; 53 HRW Photo; 54(l), Jonathan Blair/Woodfin Camp & Associates; 54(r), David R. Frazier Photolibrary; 55(b), Russ Lappa/Photo Researchers, Inc.; 55(t,c), Charles D. Winters/Photo Researchers, Inc.; 56(t,l), Walter Chandoha; 56(r), Zack Burris; 56(b), Yann Arthus-Bertrand/Corbis; 57(tc), HRW Victoria Smith/HRW Photo Victoria Smith; 57(cl), Dr. E.R. Degginger/Color-Pic, Inc.; 57(cr), Runk/Schoenberger/Grant Heilman Photography Inc.; 57(br), Joyce Photographics/Photo Researchers, Inc.; 57(l), Russ Lappa/Photo Researchers, Inc.; 57(bc), Charles D. Winters/Photo Researchers, Inc.; 58(c), Runk/Schoenberger/Grant Heilman; 59(l), Runk/ Shoenberger/Grant Heilman Photography; 59(c), Richard Megna/Fundamental Photographs; 60, Runk/Schoenberger/Grant Heilman Photography Inc.; 60, Richard Megna/Fundamental Photographs; 61(t), Kapriellan/Photo Researchers, Inc.; 62(l), Victoria Smith/HRW Photo; 63(tl), Charles D. Winters/Timeframe Photography; 63(c), Charles Winters/Photo Researchers, Inc.; 63(cr), Klaus Guldbrandsen/Science Photo Library/Photo Researchers, Inc.; 65(b), Richard Haynes/HRW Photo; 65(c), Image ©2001 PhotoDisc; 66(c), Dr. E.R. Degginger/HRW; 68(c), Michelle Bridwell/HRW Photo; 69(t), Lance Schriner/HRW Photo; 69(b), Dr. E.R. Degginger/Color-Pic, Inc.; 72(b), Victoria Smith/HRW Photo; 72(t), David R. Frazier Photolibrary; 74(t), Yann Arthus-Bertrand/Corbis; 75(t), Richard Megna/Fundamental Photographs; 76(l), Anthony Bannister/Photo Researchers, Inc.; 76(r), Richard Steedman/The Stock Market.

Chapter Four pp. 78-79 P. Loiez Cern/Science Photo Library/Photo Researchers, Inc.; 79 HRW Photo; 80(t), Dr. Mitsuo Ohtsuki/Photo Researchers, Inc.; 80(b), Nawrocki Stock Photography; 81, Corbis-Bettmann; 84, Stephen Maclone; 85(l), John Zoiner; 85(r), Mavournea Hay/HRW Photo; 87(l), Lawrence Berkeley National Lab; 88, Richard Megna/Fundamental Photographs; 91, Charles D. Winters/Timeframe Photography Inc./Photo Researchers, Inc.; 92, Superstock; 96, Nawrocki Stock Photography; 99, Fermilab National Laboratory; 100, NASA Ames ; 101(t), Stephen Maclone; 101(b), Fermi National Lab/ Corbis.

Chapter Five pp. 102-103 Jeff Goldberg/Esto; 103 HRW Photo; 109(tr), Richard Megna/Fundamental Photographs; 109(bl), Russ Lappa/Photo Researchers, Inc.; 109(br), Lester V. Bergman/Corbis-Bettman; 111(l,c), Dr. E.R. Degginger/Color-Pic, Inc.; 111(r), Richard Megna/Fundamental Photography; 112, Charles D. Winters/Photo Researchers, Inc.; 113(l,c,r), Richard Megna/Fundamental Photographs; 114(tr), Petersen/Custom Medical Stock Photo; 114(tc), Victoria Smith/HRW Photo; 115, David Parker/ Science Photo Library/Photo Researchers, Inc.; 117(t), Phillip Hayson/Photo Researchers; 118(t,c), Richard Megna/ Fundamental Photographs; 118(b), Dr. E.R. Degginger/Color-Pic, Inc.; 119(t), Michael Dalton/Fundamental Photographers; 119(b)NASA; 120 John Langford/ HRW Photo; 121 Victoria Smith/HRW Photo; 122(b), 124, Richard Megna/ Fundamental Photographs; 127, Image copyright 2001 PhotoDisc, Inc.

LabBook "LabBook Header": "L," Corbis Images, "a," Letraset Phototone, "b" and "B," HRW, "o" and "k," images ©2001 PhotoDisc/HRW; 129(c), Michelle Bridwell/ HRW Photo; 129(br), Image copyright © 2001 PhotoDisc, Inc.; 130(cl), Victoria Smith/HRW Photo; 130(bl), Stephanie Morris/HRW Photo; 131(tr), Jana Birchum/ HRW Photo; 131(b), Peter Van Steen/HRW Photo; 135, NASA; 142, Gareth Trevor/ Stone.

Appendix Page 147(t), Peter Van Steen/HRW Photo.

Sam Dudgeon/HRW Photo Pages viii-1, 5(l); 21; 31(bl); 34(r); 36; 48(b); 57(tl,tr,c); 59(r); 62(t); 63(cl); 65(cl); 66(bl); 68(t,b); 71; 73; 74(b); 75(br); 82-83; 94; 105; 108; 109(c,tl); 110; 113(b); 114(tl,b); 117(tl, br); 122(t); 128; 129(bc); 130(br,t); 131(tl); 132; 134; 136-137; 138-139; 141; 143; 147.

John Langford/HRW Photo Pages 4(t); 5(r); 6-7; 9(b); 10; 22(l); 24(all); 34(l); 63(bl,bc,br); 67(all); 129(t); 135(t).

Scott Van Osdol/HRW Photo Pages 31(br); 32(all); 33(b); 37(t); 39(b); 41-42; 133.

Self-Check Answers

Chapter 1—The Properties of Matter

Page 9: approximately 30 N

Chapter 2—States of Matter

Page 34: The pressure would increase.

Page 40: endothermic

Chapter 3—Elements, Compounds, and Mixtures

Page 59: No, the properties of pure water are the same no matter what its source is.

Page 65: Copper and silver are solutes. Gold is the solvent.

Chapter 4—Introduction to Atoms

Page 85: The particles Thomson discovered had negative charges. Because an atom has no charge, it must contain positively charged particles to cancel the negative charges.

Chapter 5—The Periodic Table

Page 114: It is easier for atoms of alkali metals to lose one electron than for atoms of alkaline-earth metals to lose two electrons. Therefore, alkali metals are more reactive than alkaline-earth metals.